Rebecca Jaymes

ROCKS & FOSSILS

ROCKS & FOSSILS

A VISUAL GUIDE

Robert R. Coenraads

FIREFLY BOOKS

A FIREFLY BOOK

Published by Firefly Books Ltd. 2005

Conceived and produced by Weldon Owen Pty. Ltd.
59 Victoria Street, McMahons Point, Sydney, NSW 2060, Australia

Copyright © 2005 Weldon Owen Inc.

First printing

Publisher Cataloging-in-Publication Data (U.S.)
Coenraads, Robert R.
    Rocks and fossils : a visual guide / Robert R. Coenraads.
[304] p. : col. ill., photos., maps ;   cm.   (Visual Guides)
Includes index.
Summary: A visual guide to how rocks, minerals and fossils provide
keys to the Earth's past.
ISBN 1-55407-068-6
1. Geology. 2. Rocks. 3. Fossils. I. Title. II. Series.
552 22   QE365.C64   2005

Library and Archives Canada Cataloguing in Publication
Coenraads, Robert Raymond, 1956-
    Rocks and fossils : a visual guide / Robert R. Coenraads.
Includes index.
ISBN 1-55407-068-6
    1. Rocks. 2. Fossils. I. Title.
QE432.C64 2005   552   C2004-907209-9

Published in the United States by
Firefly Books (U.S.) Inc.
P.O. Box 1338, Ellicott Station
Buffalo, New York 14205

Published in Canada by
Firefly Books Ltd.
66 Leek Crescent
Richmond Hill, Ontario L4B 1H1

Illustrator: James McKinnon, Wildlife Art Ltd.
Designer: Mark Thacker/Big Cat Design
Cover Design: Jacqueline Hope Raynor

Cover photo credits
Front cover top: Jesse Fisher; bottom: Photolibrary.com
Spine Jesse Fisher
Back cover (left to right): Natural History Museum; Photolibrary.com;
    Australian Picture Library/Corbis
Front flap Australian Picture Library/Corbis

Color reproduction by Chroma Graphics (Overseas) Pte. Ltd.
Printed by SNP LeeFung Printers
Printed in China

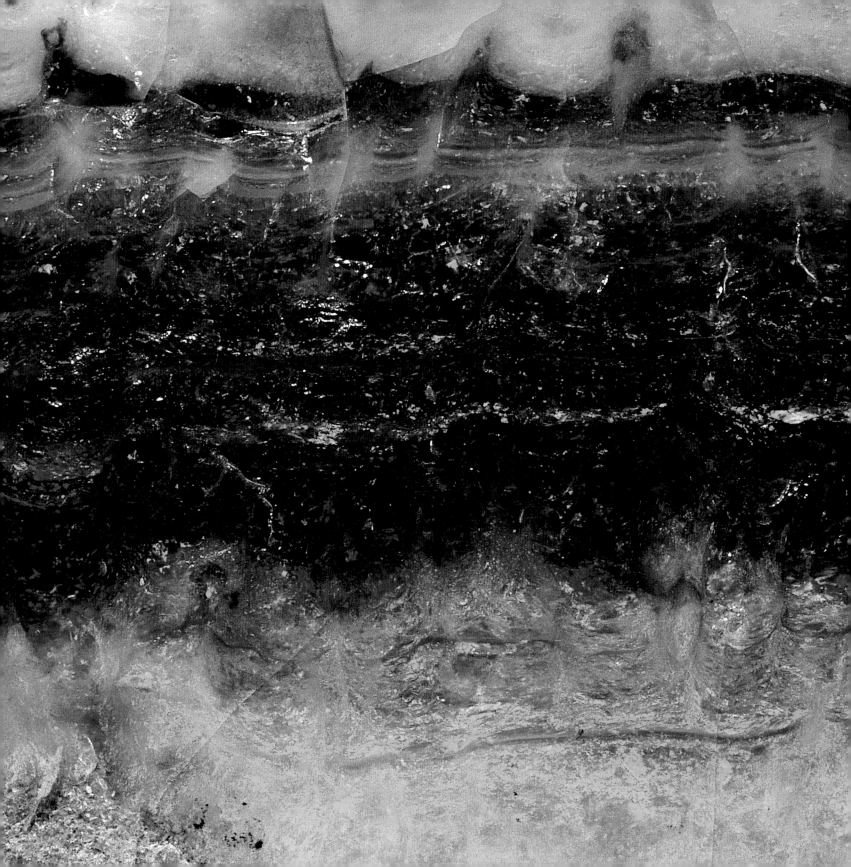

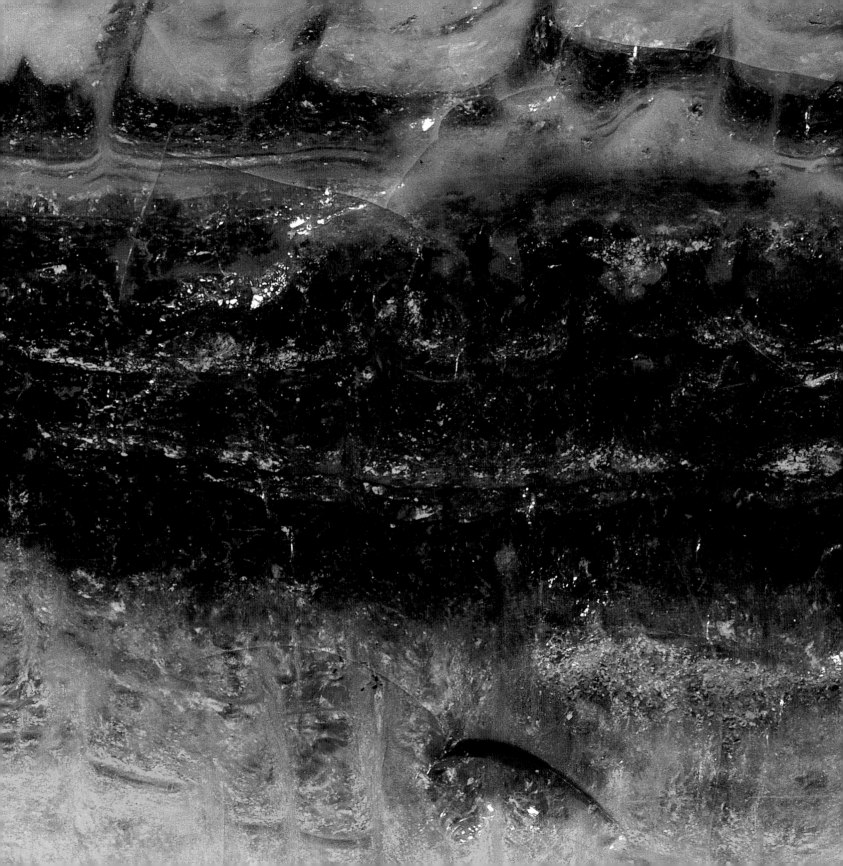

Contents

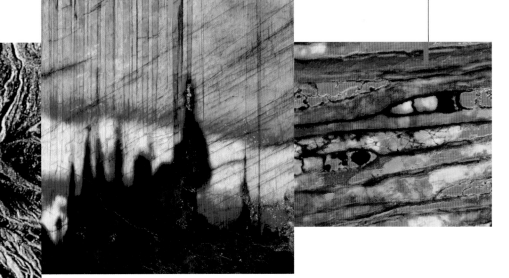

Introduction

The ground beneath us is made up of rocks, some of which extend extraordinary depths into Earth. Venture deep enough and they are molten, moving, plastic-like substances. Rocks are not static and inert. They are constantly changing, shifting and rearranging their form and location. They can be melted, deposited, eroded and squeezed into new forms. These changes occur over vast periods of time. The pages of Earth's history are written in its rocks, minerals and fossils. Rocks and minerals provide evidence of events that took place in the distant past. They offer clues about the shifting, changing nature of the continents, mountain ranges, oceans and islands. They help us understand what Earth's climate was like billions of years ago, what elements the atmosphere contained, and how life may have evolved.

When an animal or plant dies and its remains leave an impression in rock, the resulting fossil is a testament to life's history and its changing, evolving nature. Fossils reveal Earth's past lifeforms, among them long-extinct beasts, bizarre, alien-like beings and delicate marine creatures.

Rocks are made up of minerals. It is among the minerals that the exquisite gemstones and ornamental stones are found. Sparkling diamonds, vivid green emeralds, multicolored tourmalines, clear crystal quartz and lacy agates are all minerals.

Take a journey into Earth's history to discover what shaped its rocky landscapes. Delight at its sparkling minerals. Explore the fossils life has left as clues to its evolution.

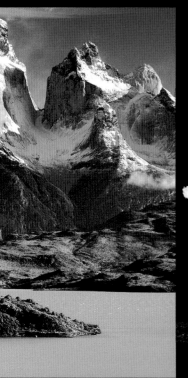

The dynamic Earth

Relentless activity driven by heat escaping from Earth's interior has shaped the oceans, divided and joined continents, and determined global climate. Life evolved to occupy the habitats presented by changing landscapes. The atmosphere formed, and as conditions changed over billions of years, so did life.

Understanding rocks

The solid part of Earth is rock. It underlies the thin, organically rich skin of air, soil and water so vital to the existence of life. Humans depend on rocks and their constituent minerals, using them for building, technology and adornment. Different rocks are found in different places—in the oceans or on the continents, in high mountain ranges or deep within Earth's interior. Rocks are continually forming, creating a diversity of landscapes around the planet. They can be explored on a grand or a small scale to reveal what they are made of and how they are classified, and to predict where certain types are likely to be found.

LANDSCAPES, ROCKS, MINERALS AND ELEMENTS

The diverse natural landscapes of the world are all made of rock. The scenery depends on the properties of foundational rocks: whether they are hard or soft; chemically stable or unstable; massive, layered or jointed; dipping or horizontal. Rocks are made of minerals, so a rock's properties reflect the nature of its constituent minerals. Common rock-forming minerals include quartz, feldspar and mica. Minerals are, in turn, made up of elements, which are comprised of protons, neutrons and electrons, the basic building blocks of planet Earth.

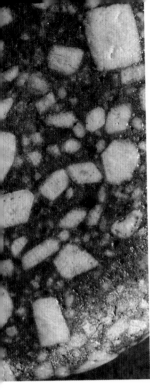

← **Simple pebbles** on a beach in Puget Sound, Washington, USA, tell a story. The product of different volcanoes located along the Rocky Mountain chain, they have been carried to the coast by glaciers, then rounded by rivers and waves. Eventually, they may become cemented into a new rock.

→ **Planet Earth** is circled by lines of volcanoes that, together with constant earthquakes, mark the edges of hard crustal plates. The active peaks, ringed by white cloud, form the island chain of Indonesia along a line where the Australian plate is being thrust below the Eurasian plate. This view was recorded by the space shuttle Atlantis in 1992 as it passed over Java.

← **Every landscape** is different, as each is formed by a unique combination of crust movement, geology, climate and time. In Monument Valley, USA, sandstones that were laid down beneath the sea during an earlier era have been uplifted and subjected to the ravages of weathering and erosion. Over time, the scenery known today was carved out. In geological time, the remaining monuments will not be around for very much longer.

Reading the rocks

From the smallest rock fragment to the largest volcano, every rock tells a story about Earth. With a little skill and knowledge, it is not difficult to read and interpret the rocky pages of Earth's history. The planet is constantly shifting, melting, eroding and changing. Geological time is vast and much change has occurred in the landscape at any given location. Indeed, through Earth's 4600 million-year history, geological forces may have moved a single piece of the crust from one pole through to the equator and out to the other pole. They may have transported it from the deepest ocean trench into the highest mountain peak or recycled it back into the mantle. Much of Earth's crust and all of its oceans have been recycled at least several times. Only small areas of tortured and folded rocks, known as continental cratons, remain from Earth's original crust.

ANCIENT ROCK FORMATIONS

Dated at more than 3 billion years old and among the oldest rocks, this banded ironstone (*right*) tells of a time when there was very little oxygen on the planet. Iron could exist then and not rust. From time to time, primitive oxygen-generating bacteria evolved and flourished, perhaps seasonally. At such times, iron in solution was readily oxidized and deposited as layers of magnetite in shallow-water basins. At times of low oxygen levels, only silica was deposited—as layers of red jasper. It was not until all the iron was oxidized and deposited that excess oxygen started to build up in the atmosphere and begin to support living things.

→ **Banded ironstone** from Western Australia shows alternating layers of gray hematite, red jasper and yellow tiger eye.

↓ **A crystalline igneous rock** collected during the Apollo 14 lunar landing in 1971 shows that Moon rocks are no different from those found on Earth.

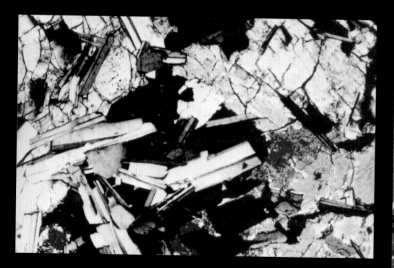

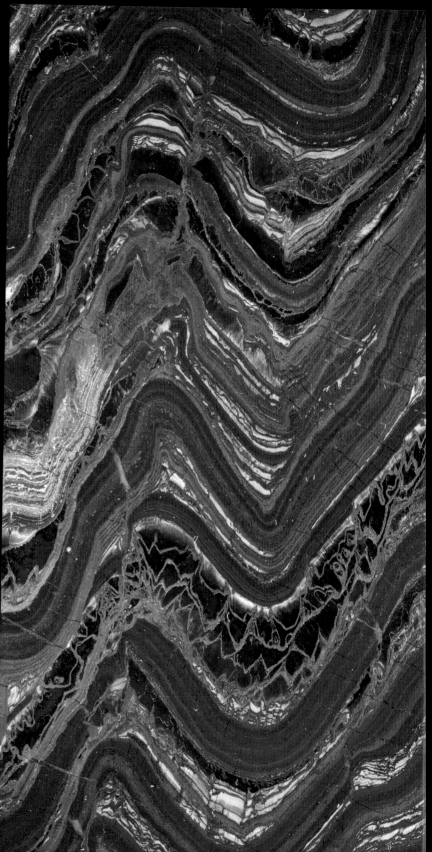

ROCKS REVEAL ANCIENT LANDSCAPES

The locations of minerals, rocks, volcanoes, mountain ranges and other landscapes are not chance occurrences. Each exists because of Earth's internal mechanics. To imagine how Earth would look without its geological engine, visit the center of an old continent. In central Australia, for example, low red desert dunes stretch from horizon to horizon. In the Amazon Basin, this vast river's tributaries meander across a swampy, forested lowland. Geological maps that show the rocks beneath the surface, however, reveal past turmoil. They show eroded volcanoes that are now marked by only a few outcrops. Vast ancient oceans have closed, leaving landlocked ridges of fossil coral limestone and deep-sea jasper. Once enormous Himalayan-style folded mountain ranges are now reduced to tiny cores of metamorphic gneiss and granulite.

↑ **Mt Everest** tells a story of seafloor rocks thrust high into the air. The sedimentary layers, highlighted by snow resting on ledges, were created by strata more resistant to erosion.

← **Tourmaline crystals** from Madagascar tell a story. Their watermelon color zoning reflects minor changes in the chemical environment as these crystal columns grew in their silica-rich host liquid.

Journey to Earth's center

To take a journey to Earth's center would be an impossible feat. It would require a vehicle capable of withstanding core temperatures of 10,000°F (5500°C) and pressures of 3,450,000 atmospheres (350 GPa). The journey would pass through several increasingly dense zones. These layers developed when Earth was still a molten ball. Under gravity, the heaviest elements migrated toward the center, displacing the lighter ones toward the outside. Traveling downward beneath the surface, the temperature increases rapidly. The average rate of increase, known as the geothermal gradient, is about 22°F per mile (25°C per kilometer). The rise in temperature is such that miners require special cooling apparatus to work the deep African diamond and gold mines. After passing through the rigid crustal rocks, the mantle is entered. Rocks here, described as rheids, have plastic properties akin to modeling clay. At temperatures high enough for these rocks to be molten, they are held in the solid state only by the immense pressure found at this depth. At Earth's outer core, temperatures are so extreme that the material is liquid. Scientists know this because certain seismic waves will not travel through liquid. Convection currents within this liquid iron zone are believed to be responsible for Earth's magnetic field. Toward the center, the core becomes firmer as increasing pressure begins to dominate, in spite of the increasing temperature. Traveling at 60 miles per hour (100 km/h), this entire journey would take about 40 hours.

INSIDE EARTH

The surface rocks of the crust are the only ones that can be studied directly. Nothing would be known about the interior of Earth were it not for the rare samples brought up from deeper places. Rocks from erupting volcanoes reveal that the mantle is largely made up of the heavy rock known as peridotite. But this reveals rocks occurring less than 100 miles (160 km) down. So scientists learn about Earth's interior by studying rocks from other planets. Several minor planets between the orbit of Mars and Jupiter that collided in the distant geological past exploded to form the asteriod belt. Fragments arriving as meteorites have revealed much about our planet's lower mantle and core. The core is pure iron.

↑ **Iron meteorites** that come from the cores of exploded planets reveal the composition and structure of Earth's core. This specimen is from Odessa Crater, USA.

→ **Olivine nodules** are pieces of Earth's mantle that are carried to the surface by volcanic eruptions. Olivine stony meteorites also exist.

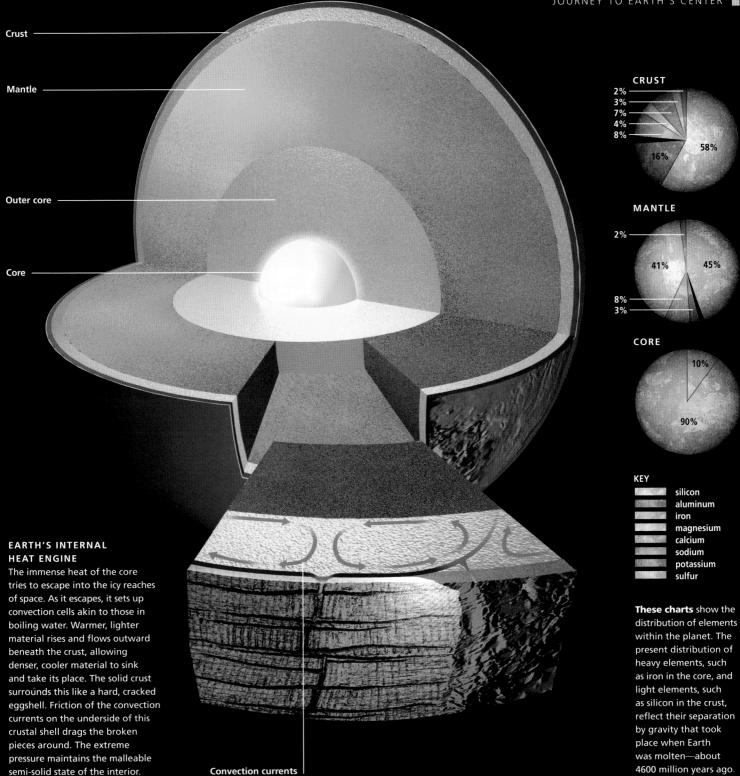

Crust

Mantle

Outer core

Core

CRUST

2%
3%
7%
4%
8%
16%
58%

MANTLE

2%
41%
45%
8%
3%

CORE

10%
90%

KEY

- silicon
- aluminum
- iron
- magnesium
- calcium
- sodium
- potassium
- sulfur

These charts show the distribution of elements within the planet. The present distribution of heavy elements, such as iron in the core, and light elements, such as silicon in the crust, reflect their separation by gravity that took place when Earth was molten—about 4600 million years ago.

EARTH'S INTERNAL HEAT ENGINE

The immense heat of the core tries to escape into the icy reaches of space. As it escapes, it sets up convection cells akin to those in boiling water. Warmer, lighter material rises and flows outward beneath the crust, allowing denser, cooler material to sink and take its place. The solid crust surrounds this like a hard, cracked eggshell. Friction of the convection currents on the underside of this crustal shell drags the broken pieces around. The extreme pressure maintains the malleable semi-solid state of the interior.

Convection currents

Rings of fire

Earth's surface is covered by nine large, rigid crustal plates plus a number of minor ones. These plates are in constant motion, riding on top of the semi-solid, plastic-like mantle. They act as an insulation blanket around Earth's hot interior. "Rings of fire" is the evocative name given to the zones marking the plate edges, a name understood best by the millions of people who live on their edges, aware of the threat of possible destruction from one of the numerous earthquakes or volcanic eruptions that occur frequently within these zones. This extraordinary concept, known as plate tectonics, was pieced together during the 20th century, and it is supported by a vast amount of observational data.

ALFRED WEGENER (1880–1930)

German geologist Alfred Wegener (*right*) originally proposed the theory of continental drift back in 1912. Impressed by the matching coastlines of South America and Africa, Wegener suggested that there had once been a supercontinent (which he called Pangea) that had fragmented and drifted apart into the continents that we know today. He was unable, however, to explain the underlying mechanism and the theory was discounted during his lifetime. Later discoveries proved him to be correct and led to the discipline of plate tectonics. Wegener died while crossing an ice sheet on an expedition to Iceland.

→ **Active volcanoes** and constant earthquake activity mark the edges of tectonic plates. Part of the Pacific rim's volcanic chain lies in the Kamchatka Peninsula in Russia. In this aerial view, snow-covered peaks punctuate the winter landscape. Mt Kronotskaya is the dominant volcano.

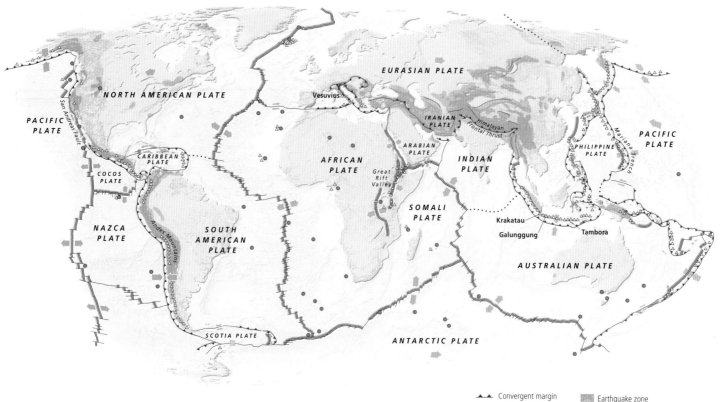

Tectonic plates are the jigsaw of crustal plates covering Earth that bump and slide against one another. Red lines indicate the edges that are moving apart while purple lines show where a plate is pushing under another (also marked by mountain belts).

- ▲▲ Convergent margin
- ══ Divergent margin
- ── Transform fault
- • • • Diffuse or uncertain
- ▰▰▰ Direction of movement
- ▦ Earthquake zone
- △ Volcanic zone
- ● Prominent hotspot

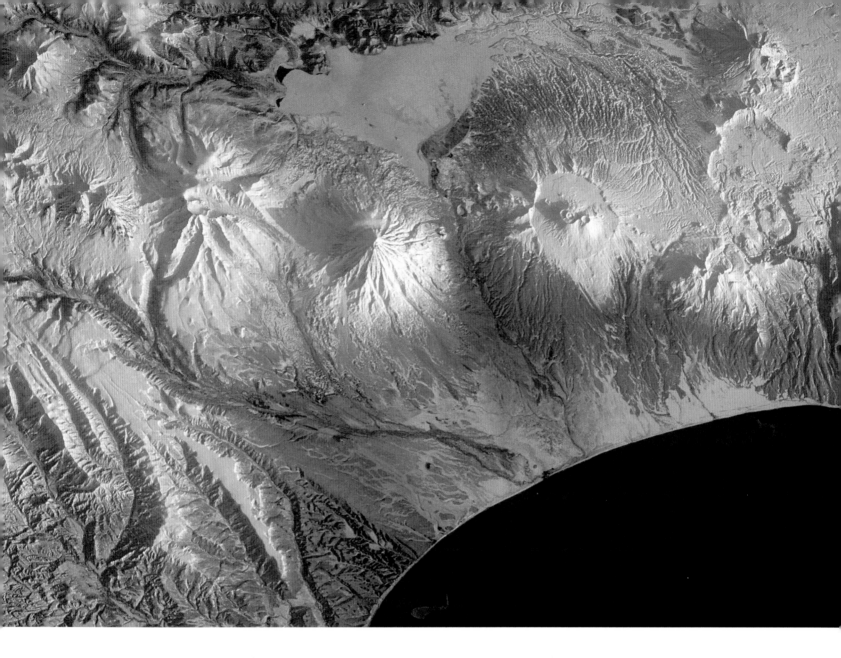

The speed at which tectonic plates move depends on the speed of the underlying convection currents that drag them along by friction. On average, this is about the rate at which fingernails grow, and is quite noticeable in any natural or built structures that cross a plate boundary.

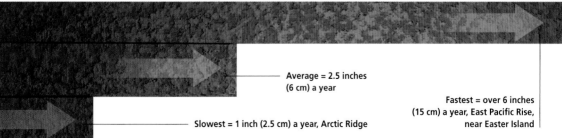

Average = 2.5 inches (6 cm) a year

Fastest = over 6 inches (15 cm) a year, East Pacific Rise, near Easter Island

Slowest = 1 inch (2.5 cm) a year, Arctic Ridge

Spreading oceans

Earth's oceans are always spreading. New crust is continually forming along the ridgelines of the oceans as they are pulled apart by forces in the mantle. For example, the Atlantic Ocean is growing. The African and American continents have slowly spread apart—on a map it is clear how perfectly they fit together (first noted by Alfred Wegener) and the geology on either side matches. But Earth is not increasing in size, so the growing oceans must be balanced by crustal consumption elsewhere. This is occurring in the Pacific. A dying ocean, its floor is being consumed in the trenches around its edges, known as the Pacific "ring of fire." Ocean floor basalt can be dated, and these ages record a symmetrical aging on either side of the central ridgeline. Compared to the continents, the oceans are unusually young because the tectonic plate processes cause oceans to be continuously born and consumed. Being lighter, the continents resist being consumed, like the foam that accumulates around the edges of a saucepan of boiling soup. Instead, they become squeezed together in parallel rows of fold belts.

Huge fissures mark the line of the mid-Atlantic spreading ridge on which Iceland sits. This small, volcanically active island straddles the American and Eurasian plates, which are moving apart at a rate of about 1.2 inches (3 cm) per year.

Viewed from the orbiting Gemini II spacecraft, the Red Sea and Gulf of Aden are the beginning of a completely new ocean as the Arabian and African–Somali jigsaw pieces separate from one another.

HOW AN OCEAN IS BORN

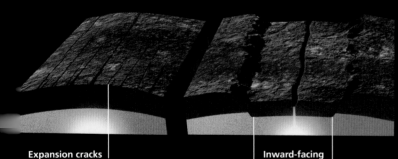

Expansion cracks

Inward-facing faults

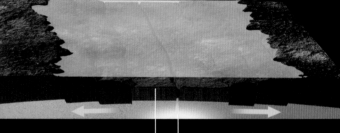

New seafloor | Upwelling basalt

1. CRACKING

Strong upwelling currents beneath the continents cause local uplift and doming, much like water boiling in a lidded saucepan. Rivers flow outward from these areas and tension cracks slowly begin to develop.

2. UPLIFT AND COLLAPSE

As uplift, expansion and cracking continue, the central block collapses along inwardly facing fault planes, forming a straight-edged rift valley. Rivers may localize in these valleys, but the major drainage continues to flow away from this uplifted zone. The East African Rift is an example of such a valley.

3. CONTINUED RIFTING

Separation of the two continental masses leaves space for upwelling basaltic magmas to intrude along the cracks and erupt onto the rift valley floor. As the cracks grow, basalt continues to fill them. Finally, when separation is sufficient and the crust sinks, ocean waters flood into the rift valley. At first, this process is slow and evaporation of the seawater allows huge salt deposits to form in semi-permanent lakes. Eventually, the water remains in the valley and a new ocean is born. The Red Sea and Gulf of Aden are examples of the formation of an ocean.

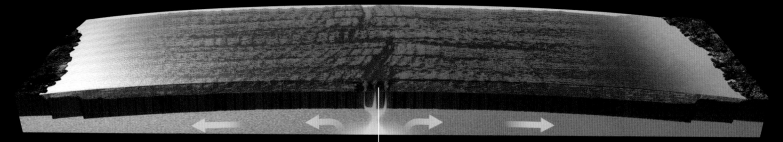

Mid-oceanic ridge

4. MATURE OCEAN

Spreading continues as new basalt is added along the mid-oceanic centerline. Because it is hotter and above the cell, the mid-oceanic ridge is more buoyant and shallower than elsewhere, but it maintains its rift valley. The cooler ocean floor deepens progressively away from the spreading ridge and reaches maximum depth adjacent to the continental margins. This process continues as long as the convection currents maintain their strength. Eventually, currents beneath another large continent will open a new ocean, forcing the old one to close.

Subduction

Subduction is the sliding of one crustal plate beneath another. It occurs as a consequence of rifting and the creation of new ocean floor elsewhere on the globe. Earthquakes and mountain ranges of active volcanoes that lie adjacent to deep oceanic trenches mark the path of the descending plate as it plunges below the other plate. The locations of earthquakes reveal a planar zone known as the Wadati–Benioff zone, named for the seismologists who discovered it. This zone starts in the trenches and dips at a steep angle away from the ocean basins. Most earthquake activity occurs at depths shallower than about 40 miles (70 km) as the cold, brittle descending plate pushes and grinds its way against the other plate. As this plate descends, earthquake activity lessens because heating means the rocks bend like plastic rather than cracking. Melting of the upper surface of the descending slab begins, aided by the limy, salty, water-rich, ocean-floor sediments riding on top of it. The resultant andesitic magmas rise to the surface, erupting violently to produce stratovolcanoes. Below about 180 miles (300 km), earthquakes are restricted to the interior of the cool, descending slab until it reaches a depth of about 440 miles (700 km), when the slab's interior finally reaches equilibrium with the surrounding hotter rocks.

Andesitic lava rises from the melting Cocos plate as it is pushed beneath the Caribbean plate, causing the volcano Arenal to erupt violently, ejecting ash and lava. Located in Costa Rica, Arenal is one of the numerous active subduction-related stratovolcanoes that ring the Pacific Ocean. They mark the edges of this ocean's destruction in the subduction trenches.

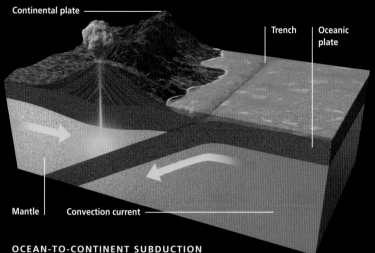

Continental plate — Trench | Oceanic plate

Mantle | Convection current —

OCEAN-TO-CONTINENT SUBDUCTION

When oceanic crust collides with continental crust, the heavier basaltic oceanic crustal plate is overridden and slides below the lighter, thicker granitic continental crust. The subducting crust and overlying sediments melt, producing a chain of explosive volcanoes. Ocean-to-continent subduction is seen along the west coast of South America with the classic Andean cordillera reaching heights of 22,835 feet (6965 m) at Aconcagua, Argentina. It is immediately adjacent to the Peru–Chile Trench, which reaches depths of 26,455 feet (8069 m).

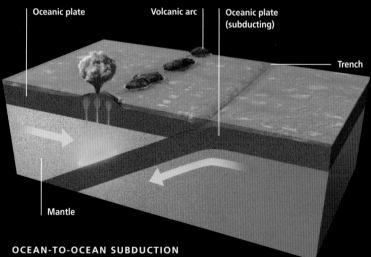

Oceanic plate | Volcanic arc | Oceanic plate (subducting)

Trench

Mantle

OCEAN-TO-OCEAN SUBDUCTION

When two oceanic crustal plates collide, one must slide beneath the other, usually the older, colder one as it is slightly denser. Magmas rising from the descending slab as it melts erupt onto the ocean floor, forming volcanoes that build up above sea level and eventually coalesce to form an island archipelago adjacent to a deep trench. Classic examples include Indonesia and the Java Trench, Japan and the Japan–Kurile Trench, the Aleutian Islands and the Aleutian Trench.

The subduction volcano Gunung Semeru (*above*) lies in the Indonesian island archipelago and was formed by ocean-to-ocean subduction, while the volcano peaks of Torres del Paine National Park (*below*) in Patagonia, Chile, were formed by ocean-to-continent subduction. Both areas are considered to be active, but the Torres del Paine peaks have been glacially eroded since the last eruption.

Collision

Continental collision is a type of subduction that occurs when an ocean closes and the continental landmasses on either side collide with one another. Neither landmass is able to descend, being too light and bulky, so they just continue to push against one another, crushing, folding and deforming the rock. Seafloor sediments that become trapped between the closing continents are folded into linear belts, metamorphosed and pushed up into tall mountain ranges. In the Himalayas, this is evidenced by seafloor sandstones and limestones found on top of Mt Everest at 29,035 feet (8856 m) above sea level. Eventually, the colliding continents cause the subduction process to jam and stop. In order to relieve the immense stress that is constantly building up, the subduction zone then "jumps" to a new location where normal oceanic crust can continue to disappear unimpeded. While this process merges continents into larger landmasses, the heat trapped beneath can build up and restart the process of rifting, giving birth to a new seafloor.

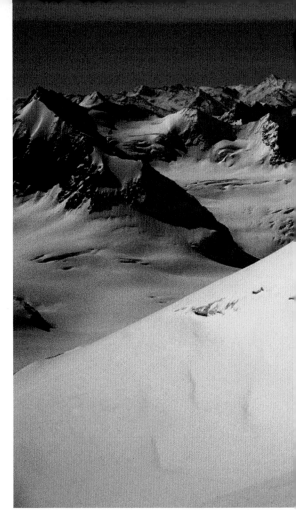

CLOSING OF THE TETHYS SEA

About 250 million years ago, all the continents were joined in a single supercontinent, Pangea. About 180 million years ago, it began to break up, slowly enclosing the Tethys Sea. Accelerated by the splitting of Africa and South America, the African–Arabian and Indian plates traveled northward, swallowing the Tethys seafloor until they slammed into the Eurasian plate. This created a collision mountain belt some 6000 miles (10,000 km) long that is still being pushed upward. All that remains of the once-mighty Tethys is a number of small, semi-restricted seas and lakes, including the Caspian and Black seas.

→ **The Swiss Alps** form part of the huge collision mountain belt that resulted from the closure of the Tethys Sea. The belt includes the Atlas Mountains, Alps, Dinaric Alps, Pindus Mountains, Zagros Mountains, Hindu Kush and Himalayas. All are characterized by numerous earthquakes and highly folded rocks.

Satellite view of the folds in the Atlas Mountains of Morocco. These mark the collision of the African and Eurasian plates, which closed the Tethys Sea.

Microscopic folds in schist are the product of squeezing and metamorphosing organically rich sandy and clayey seafloor sediments.

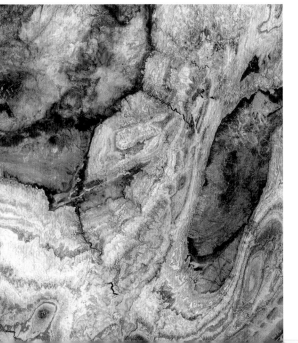

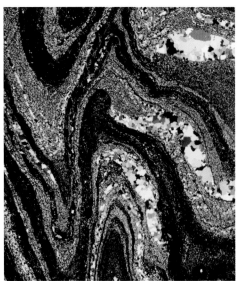

HOW THE HIMALAYAS FORMED

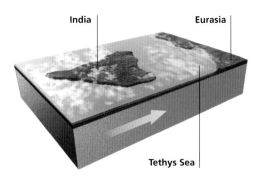

India | Eurasia

Tethys Sea

1. INDIA STARTS ITS JOURNEY

When Pangea broke up, India was part of the great southern continent, Gondwana, which in turn began to split. India broke away about 145 million years ago and started its northward journey toward Eurasia, reducing the size of the Tethys Sea.

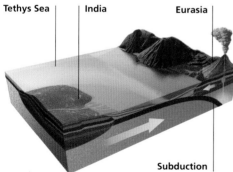

2. OCEAN-TO-CONTINENT SUBDUCTION

As India approached Eurasia, its seafloor subducted beneath the Eurasian continental landmass. A line of andesitic volcanoes developed along the edge of the Eurasian plate adjacent to the trench, much like those that occur along the coast of Chile.

Labels: **Tethys Sea**, **India**, **Eurasia**, **Subduction**

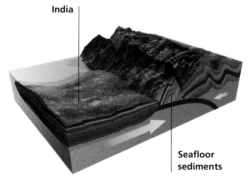

3. COLLISION BEGINS

Eventually India was pushed against Eurasia as the last remnants of the Tethys Sea were subducted. Seafloor sediments trapped between the continents were compressed and pushed upward, marking the birth of a tall mountain range—the Himalayas.

Labels: **India**, **Seafloor sediments**

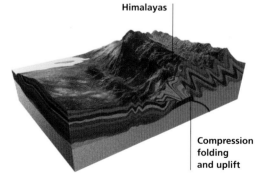

4. FOLDING AND UPLIFT

Compression continued, turning the sediment into metamorphic rock, squeezing it upward and outward along thrust faults over the adjacent continents, a process which continues today. As India cannot subduct, the process will eventually jam and stop.

Labels: **Himalayas**, **Compression folding and uplift**

Earthquakes and faults

At transform margins, crust is neither created nor destroyed as plate motion is neither divergent nor convergent. The plates slide beside one another in opposite directions along a strike-slip or lateral fault. Plate motion is slow, but the energy released can be immense as the fault moves in short, sudden steps. Careful monitoring of earthquake timing and location along the San Andreas Fault over decades has enabled scientists to calculate which portion of the fault has stress building and where it has been relieved by earthquakes. While this semi-predictive approach can highlight dangerous areas of the fault, it unfortunately cannot predict the date and time when the next "big one" will occur.

TSUNAMIS

Displacement of the seafloor will generate tsumanis—waves traveling outward in all directions from the source. In deep open ocean, they can reach speeds of 300 to 600 miles per hour (480 to 960 km/h), but can pass practically unnoticed as their apparent height is under 3 feet (1 m). The Pacific Tsunami Warning System is a series of gauges noting abnormal changes in sea level. It enables populated coastlines at risk to be given up to 20 hours warning of an approaching tsunami.

→ **A tsunami** is about to kill this man as it sweeps over a pier in Hilo, Hawaii. He was one of 159 people killed by a wall of water set off by a 7.3-magnitude earthquake in the Aleutian Islands in 1946.

I   II   III   IV   V   VI   VII

Strike-slip faulting is most clearly visible from the air. Outcropping geological strata in Iran are offset by horizontal displacement along the fault line. A sudden movement, even a small one, is enough to generate large amounts of earthquake energy.

THE MERCALLI SCALE

The Mercalli scale is a way of measuring earthquake energy arriving at a given point by measuring the actual damage and also using information from observers. The better-known Richter scale measures the absolute magnitude gleaned from seismograph data, regardless of the damage caused by the earthquake. Generally, the effects of an earthquake decrease radially away from the epicenter unless special conditions, such as soft, wet sediments, amplify the energy. This is what occurred in the 1985 Mexico City earthquake, a city built on an old lake.

Mercalli intensity

I. People feel no movement. **II.** Those on upper floors of tall buildings may notice movement. **III.** Many people indoors feel movement. Hanging objects swing. Vibration like a passing truck. **IV.** Dishes, windows and doors rattle, parked cars rock. People outdoors may notice earthquake. **V.** Sleeping people wake. Doors swing. Dishes break. Small objects move. **VI.** Everyone feels movement, many run outdoors. Walking is difficult. Furniture moves. Pictures fall. Trees shake. Slight damage in poorly built structures. **VII.** Standing is difficult. Cars shake. Loose bricks fall. Damage slight in well-built buildings, bad in poor buildings. Some chimneys break. **VIII.** Steering cars becomes difficult. Chimneys fall. Tree branches break. Heavy furniture overturns. **IX.** Well-built buildings considerably damaged. Ground cracks, pipes break. **X.** Most buildings and foundations destroyed. Dams seriously damaged. Large ground cracks and landslides. Railroad tracks bend slightly. **XI.** Most buildings collapse. Some bridges are destroyed. Underground pipes destroyed. **XII.** Total damage. Ground moves in waves. Objects thrown in air.

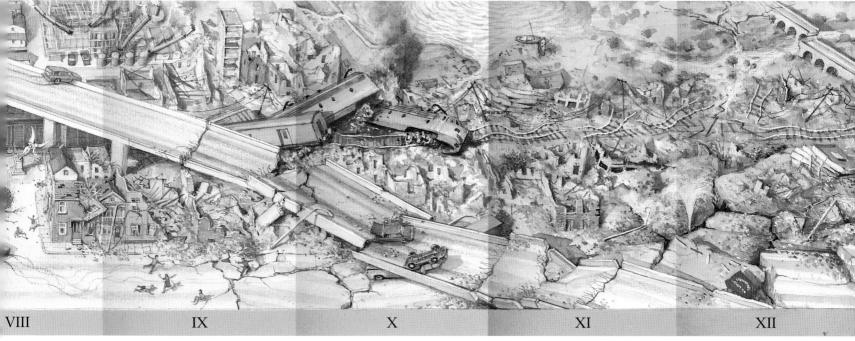

VIII   IX   X   XI   XII

Hot spots

Hot spots are rising plumes of hot, buoyant magma that originate somewhere in the mantle and periodically reach the surface through weaknesses in the overlying, moving crustal plates. These mantle plumes are geographically fixed with respect to one another within the planet. Over time, a hot spot will produce a line of volcanoes as the overlying crustal plate moves over it, much like a sewing machine needle punching holes in the fabric as it passes beneath. Each volcano lasts a short time, perhaps only a few million years, becoming inactive as it is shifted away from the rising plume. This is most spectacularly seen in the numerous parallel chains of volcanic islands in the Pacific Ocean, such as the Hawaiian, Tahitian and Christmas island chains. They follow a roughly southeasterly-to-northwesterly direction (that is, at right-angles to the East Pacific spreading ridge), each with an active volcano or seamount at its southeasterly end. Unlike other volcanic activity, which is related to plate boundaries, hot spots apparently occur randomly within the plates, both in ocean basins (where they are most easily seen) and beneath continents (such as the Yellowstone hot spot). The reason for this is unknown.

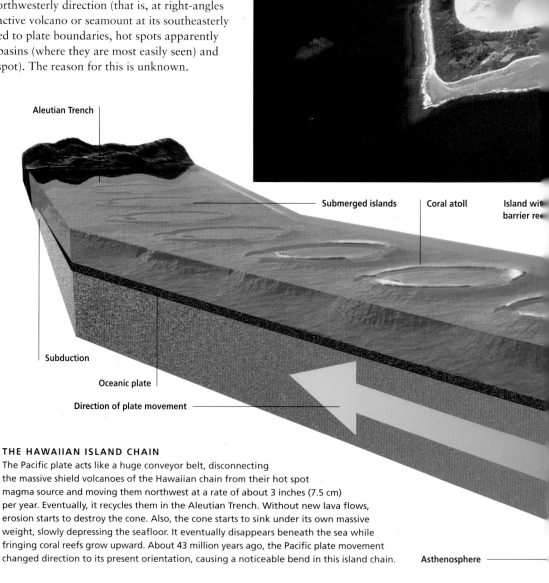

ACTIVE ISLANDS

On Hawaii's Kilauea volcano, lava spouts skyward from a small vent (*below*). Such active volcanic islands mark the position of a hot spot. Cooling and solidifying in the air, black ash and cinders fall to the ground, building up a cone around the vent. Prevalent winds have caused the cone to grow asymmetrically. Hawaii's current hot spot is the big island of Hawaii. It is becoming less active as the island has started to move away. A new volcano, Loihi, is growing on the seafloor to the southeast and is expected to break through to the surface in about a million years.

Aleutian Trench

Submerged islands

Coral atoll

Island with barrier reef

Subduction

Oceanic plate

Direction of plate movement

Asthenosphere

THE HAWAIIAN ISLAND CHAIN

The Pacific plate acts like a huge conveyor belt, disconnecting the massive shield volcanoes of the Hawaiian chain from their hot spot magma source and moving them northwest at a rate of about 3 inches (7.5 cm) per year. Eventually, it recycles them in the Aleutian Trench. Without new lava flows, erosion starts to destroy the cone. Also, the cone starts to sink under its own massive weight, slowly depressing the seafloor. It eventually disappears beneath the sea while fringing coral reefs grow upward. About 43 million years ago, the Pacific plate movement changed direction to its present orientation, causing a noticeable bend in this island chain.

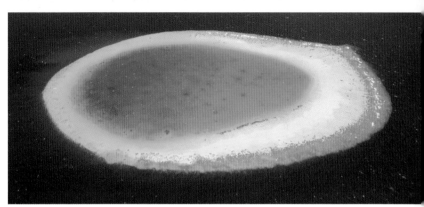

Bora-Bora Island lies in the Tahitian chain. Only the deeply eroded core of the original shield volcano remains, ringed by a lagoon and protective barrier reef. As the island subsides, the reef grows upward, marking the island's former edge.

Opposite ends of an island chain are represented above. Fresh cinder cones in Haleakala caldera, Hawaii (*top*) are near the hot spot end. North Male coral atoll, Maldives (*above*), is perched above a sunken volcano at the far end of a chain.

Extinct volcano with fringing reef

Active shield volcano

Underwater seamount

Understanding fossils

Fossils are found throughout Earth's sedimentary rocks and they help geologists to determine the geological period and environment in which those rocks were deposited. Earth's rock strata can be read like a giant, stony, history book of life throughout geological time. Fossils enable the lineages of today's organisms to be compiled into a detailed evolutionary chronology. There are many gaps or distortions in the record as organisms with hard body parts, such as shells, skeletons and teeth, are more readily fossilized than those that are soft bodied, and water dwellers are more commonly preserved than those living on land. Delicate organisms, too, such as the brittle star below, can be preserved only under special circumstances. The book of life makes exciting reading—there are strange Cambrian-dawn era creatures, missing links, dead-end lines, worldwide calamities and periodic devastating mass extinctions wiping out almost all lifeforms. It tells about the rise of more complex forms, culminating in an intelligence capable of reviewing and understanding its own evolutionary process.

↙ **Brittle stars** have changed little since they evolved in the Ordovician. Here, *Ophioderma*, an early Jurassic fossil brittle star, can be compared to its modern counterpart (*below*), foraging on a red sponge.

→ ***Seymouria*** is a key evolutionary linking fossil, between the amphibians and emerging reptiles. It lived during the early Permian, about 280 million years ago. It had an amphibian-like skull with a reptilian body form and scaly skin.

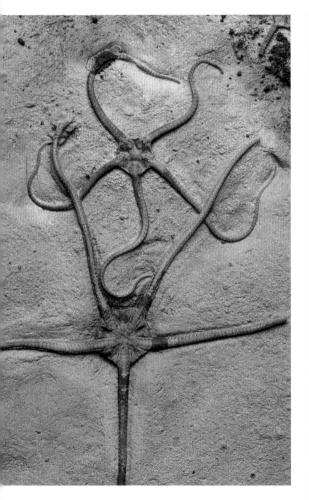

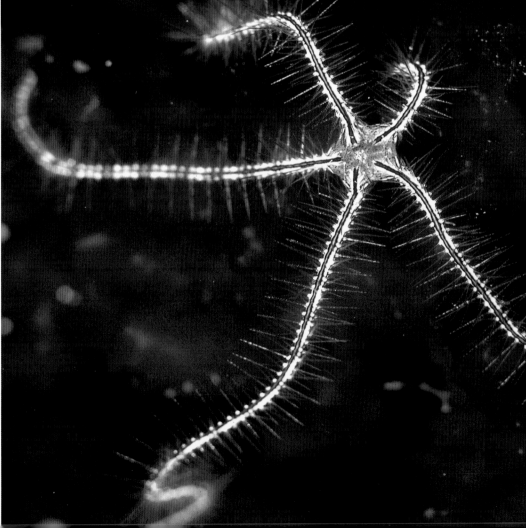

EVOLUTIONARY LINKS

All fossils have a story to tell and paleontologists can piece the evidence together like a jigsaw puzzle to reconstruct a continuous record of life from its humble beginnings to the present time. Particularly important are those fossils that mark the points where species evolve into new lines as they adapt to niches that become vacant, or changing environmental conditions. These linking fossils (such as *Seymouria*, at right) are often rare as the adaptive changes to certain opportunities appear to occur quite quickly in geological time, as evidenced by the Galapagos finches. Similar adaptive changes led to the rise of the birds, mammals and humans.

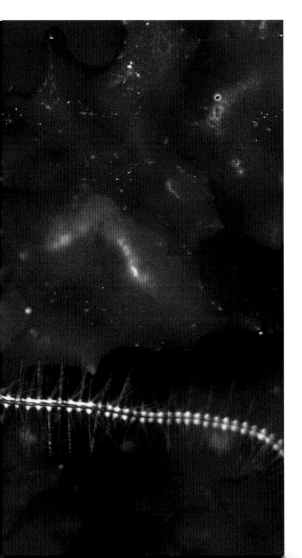

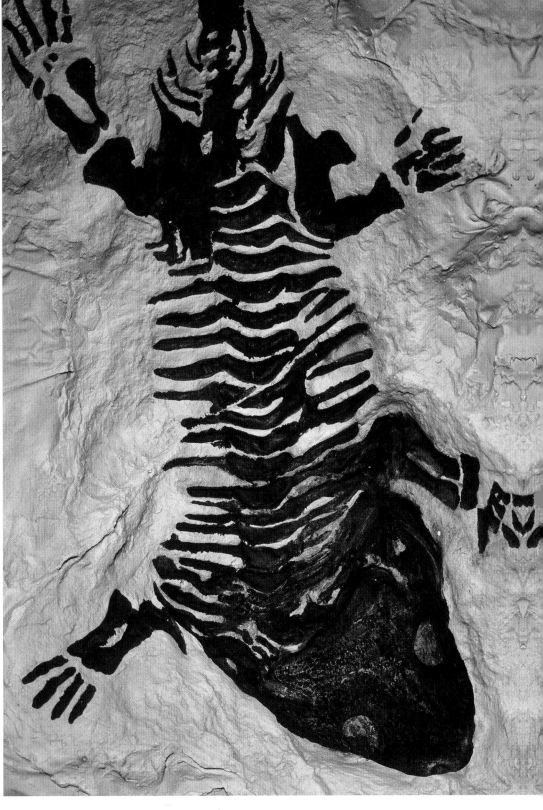

The fossil record

Planet Earth began to support life about 3.5 billion years ago, and from these simple organisms evolved increasingly more complex lifeforms. It is only through the discovery of fossils, which record much of this development, that we understand how life has changed over such enormous stretches of time. Fossils can vary in size from invisible specks of microfossils to gigantic communal structures, such as reefs, that now form vast mountain ranges. Each fossil, small or grand, reveals a piece of a story that is still evolving. Fossils can tell real-life stories of moving continents, ancient oceans, mountain ranges, strange life and unimaginable global catastrophes.

FOSSILS AND PLATE TECTONICS

Fossils track the movement of continents between the equator and the poles by recording the changing climate. Regional uniformity of life in the fossil record reveals when continents were joined, while divergences and increases in species indicate when the continents were separated and life evolved independently.

→ **Microfossils of foraminifera,** *Nummulites* species, in this limestone reveal that they were laid down in a 50 million-year-old Eocene deep seafloor.

↓ ***Albertosaurus*** **is a carnivorous dinosaur** from Canada. Such fossils tell of long-extinct creatures that once roamed the planet. The dinosaurs perished 65 million years ago in one of many planet-wide catastrophes.

THE WORLD'S LARGEST SPONGE REEF

The largest bioconstruction known on Earth is a reef built by calcareous sponges 145 million years ago. This reef grew in the shallow, warm shelf waters at the edge of the ancient Tethys Sea, which teemed with now-extinct ancient life, including ammonites and plesiosaurs. Apart from a few small remnant lakes and seas in Europe, the Tethys Sea has long since closed and disappeared. A massive belt of limestone stretching across Europe is all that remains to mark the position of this gigantic structure. These limestone rocks provide evidence of living communities and climatic conditions that were completely different from those of Europe today. The hard-working reef-building sponges were also thought to be extinct until a living community was recently discovered on the ocean floor off the west coast of Canada. Many enormous reef structures made by different calcium-carbonate-producing organisms have existed through time. These structures show reef-building conditions far more prolific than today. Although massive, a living reef is highly sensitive to global climatic change and the effects of environmental pollution.

↑ **The Dolomite Mountains**, Italy, are limestone and dolomite beds that formed in an ancient ocean. They were uplifted by the movement of tectonic plates and sculpted by slow-moving glaciers. Today, exposed on their steep faces are enormous cross-sections of a Triassic coral reef community.

The five kingdoms

Like all living organisms, fossils are classified into major groupings known as kingdoms. This helps scientists to study and understand how life evolved and even where it may eventually head. Originally, only two kingdoms—plants and animals—were recognized. It became clear that further divisions were needed to include organisms such as fungi and bacteria. There were also microscopic organisms that display unusual, mixed characteristics that fall between several of the other kingdoms. The five-kingdom system was developed in 1963, with the other three kingdoms being the Monera (primitive, non-nucleated bacterial cells); Protoctista (single-celled organisms); and Fungi (multicelled organisms). Protoctista contains such a diversity of lifeforms that many consider are different enough to warrant classification into several further kingdoms. Fossils from all kingdoms are abundant, with Monera fossils appearing 3900 million years ago; Protoctista, 1700 million years ago; followed by Animalia (600 million years ago); Fungi (550 million years ago); and Plantae (500 million years ago).

THE TREE OF LIFE'S KINGDOMS

The roots of the tree of life are buried in Earth's prehistory. At the very bottom are the Monera, tiny single-celled organisms without a nucleus. These cells are called prokaryotes and include the ever-present bacteria. Protoctista were the first single-celled organisms to develop a nucleus, and because of this, they are classified as eucaryotes. These include plankton and amoebas. Protists form the trunk for the other three kingdoms that started branching during the Cambrian. These are: animals, which obtain their life energy through consumption of other life (heterotrophs); plants, organisms that produce their own food using the energy from the Sun (autotrophs); and fungi, which feed by breaking down dead or decaying material and absorbing the nutrients (saprophytes).

→ **Kingdom Fungi,** such as this toxic fly agaric mushroom, consists of life that feeds on decomposed matter.

→→ **Kingdom Protoctista** contains single-celled lifeforms. Slime mold is a giant single cell with multiple nucleii and can slowly move to feed.

↘ **Kingdom Animalia** includes multicellular organisms and ancient animals, such as this Jurassic crayfish.

↑ **Kingdom Plantae** includes mosses, ferns, algae, conifers and flowering plants; all use chlorophyll and sunlight to photosynthesize their own food.

↑ **Kingdom Monera** contains the earliest known single-celled life on Earth. This 10 million-year-old fossil bacterial cell was found in seafloor chert.

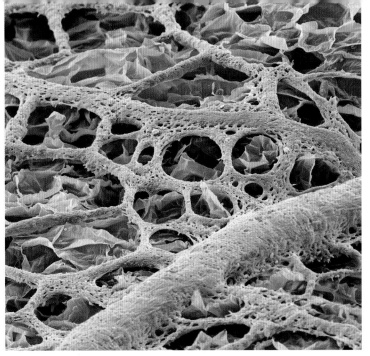

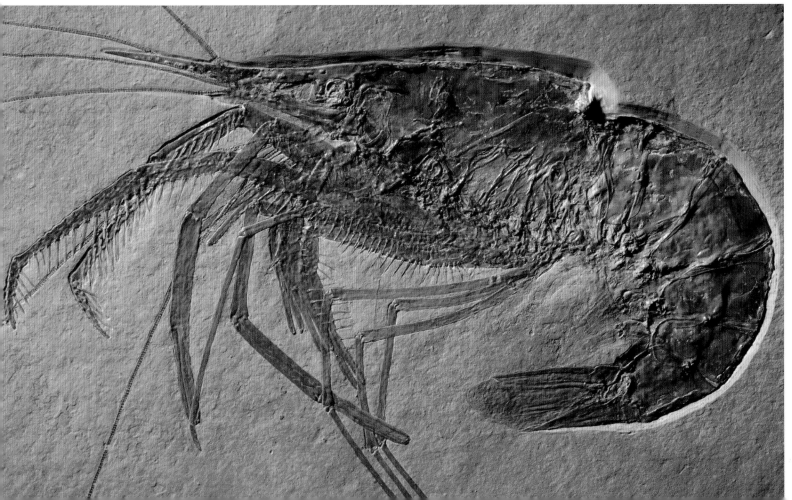

Classification of fossils

Once placed into its correct kingdom, every lifeform, fossil or living, can be then classified into a successively finer series of categories using a universally applicable naming system. The smallest category in the system is the species, members of which may reproduce sexually with one another (something that is difficult to test with fossils). Species are grouped into genera (plural of genus), which are grouped into families. Families, in turn, are grouped into orders, classes, phyla (plural of phylum) and kingdoms. This system was developed by Carl Linnaeus and published in 1757. The universal convention is that the two-part name (or binomial) is always written in italics or underlined with the name of the genus beginning with a capital letter. The generic name may be used alone (for example, *Felis*) but the species name must always be written with the genus (*Felis catus*). After being written in full the first time, the genus is abbreviated to the first letter (*F. catus*). Depending on preservation, fossils may be difficult to classify. Further difficulties can arise if juveniles, males and females look different to one another. This can sometimes lead to incorrect classification.

BIODIVERSITY
Biological diversity or biodiversity is the number and genetic variety of organisms living in Earth's giant interdependent ecosystem. Scientists reasonably guess that there may be about 10 to 20 million species with as few as 1.4 million described. Biodiversity is greatest around the equator, with tropical forests containing vast numbers of species. Despite this, more than 99.9 percent of all species that have lived since the beginning of life are now extinct.

An astounding variety of species is found in coral reefs and tropical rain forests. More than 90 percent of these remain to be identified. The main threat to biodiversity comes from humans.

This **Oligocene weevil** belongs to the order Coleoptera (beetles) of the class Insecta. Forty percent of all animals are beetles (more than 350,000 species), making it the largest order in the animal kingdom.

Carl Linnaeus, here just avoiding death on an expedition in Finland, devised the classification system of life, called the Linnaean Taxonomic System.

LINNAEAN CLASSIFICATION SYSTEM

Classification of the domestic cat (*right*) shows how the Linnaean system of nested categories works—each one contains organisms with progressively similar characteristics. In the kingdom Animalia, phylum Chordata includes all those animals with a centralized nerve chord. Subphylum Vertebrata have the spinal chord in a column of bony vertebrae connected to the head. Class Mammalia contains vertebrates that are warm-blooded with hair, milk glands and a four-chambered heart. Subclass Eutheria are those mammals with a placenta that give birth to live young. Order Carnivora have specialized teeth for meat eating. Family Felidae contains all of the cats, both large and small. Genus *Felis* is a grouping of all species of small cats. Finally, the species, commonly called the domestic cat, is known scientifically by the two-part name of *Felis catus*.

SPECIES: *catus*—domestic cat

GENUS: *Felis*—domestic cat, sand cat, jungle cat, black-footed cat, Chinese desert cat

FAMILY: Felidae—domestic cat, lion, tiger, leopard, panther, puma, lynx, *Smilodon*

ORDER: Carnivora—domestic cat, seal, wolf, dog, bear

CLASS: Mammalia—domestic cat, human, lemur, dolphin, platypus, woolly mammoth

PHYLUM: Chordata—domestic cat, fish, salamander, dinosaur, albatross

KINGDOM: Animalia—domestic cat, stick insect, sea urchin, sponge

Evolution

Through astute observation of life in the wild during the mid-nineteenth century, both Charles Darwin and Alfred Wallace independently arrived at an explanation of how life has evolved through the process of natural selection or survival of the fittest. This encompasses the idea that those members of any species better adapted to a certain habitat—being stronger, bigger, more agile, more aggressive or more intelligent—are more likely to survive and pass their superior genes on to the next generation. As advantages for survival in one environment may be disadvantageous in another, this leads to speciation, or the evolution of specific characteristics in different parts of the world. Consider how the process of evolution works in Wallace's own words from 1858: "Neither did the giraffe acquire its long neck by desiring to reach the foliage of the more lofty shrubs, and constantly stretching its neck for that purpose," but because "those with a longer neck than usual at once secured a fresh range of pasture over the same ground as their shorter-necked companions, and at the first scarcity of food were thereby enabled to outlive them."

↘ **The bottlenose dolphin** is a highly efficient marine mammal. The basic body shape and method of movement is convergent with that of ichthyosaurs.

→ **Kronosaurus**, a pliosaur or meat-eating marine reptile, evolved flippers and returned to the sea in order to take advantage of abundant food supplies. Pliosaurs became extinct during the Cretaceous.

↓ **Three big cats** of the genus *Panthera* evolved separately on different continents: *Panthera leo*, the lion, in Africa; *Panthera onca*, the jaguar, in South America; and *Panthera tigris*, the tiger, in Asia and India. This digitally composed image illustrates the results of divergent evolution.

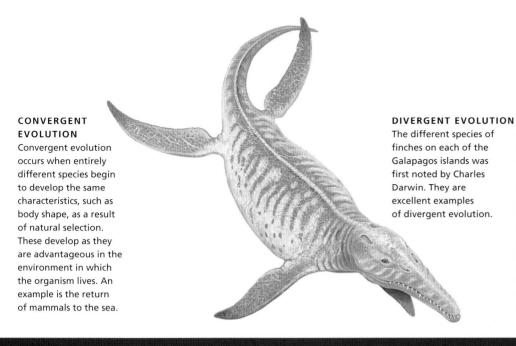

CONVERGENT EVOLUTION

Convergent evolution occurs when entirely different species begin to develop the same characteristics, such as body shape, as a result of natural selection. These develop as they are advantageous in the environment in which the organism lives. An example is the return of mammals to the sea.

DIVERGENT EVOLUTION

The different species of finches on each of the Galapagos islands was first noted by Charles Darwin. They are excellent examples of divergent evolution.

After finches arrived on the islands from South America some 10,000 years ago, their beak shape and size changed as they adapted for grasping, probing, crushing or cutting according to the different food sources. Foods include flowers, seeds, ticks, eggs, insects and even the blood of other birds.

Vampire finch—eats blood

Tree finch—eats insects

Warbler finch—eats insects

Woodpecker—eats insects

Tree finch—eats plants

Cactus ground finch— eats cactus

Ground finch—eats seeds

Adaptation

As part of the process of evolution, a species must adapt to the conditions in which it finds itself in order to reproduce. Such conditions include excesses of light, dark, heat, cold, dryness, salt or predation by others. For example, desert animals such as the oryx do not sweat; similarly, desert plants have fewer, smaller pores to reduce water loss. The pronghorn antelope has large muscles and lungs to maintain speeds of 40 miles per hour (60 km/h) to avoid wolves. Electric fishes and eels use battery organs and electroreceptors to communicate and navigate in dark waters. Fossil evidence indicates that adaptation has been occurring since life began.

→ **An elephant** in the Andaman Islands, India, enjoys a swim to cool down. Another way it reduces heat is to pump blood through numerous small capillaries in its large, thin ears. Elephants can also flap their ears to aid this process as well as to ward off insects.

↓ **The Arctic fox,** *Alopex lagopus*, has a number of highly specialized adaptations to extreme seasonal changes. In winter its fur becomes thicker and changes from gray to white, affording both warmth and camouflage. Fur around its footpad also protects against the cold and prevents slipping on the ice.

ADAPTING TO HEAT AND COLD

Many adaptations are aids for survival in extreme conditions. For the cold, these include thick fur, layers of body fat and antifreeze blood of coldwater fishes and frogs. In hot climates, some plants require fire to set seed while certain animals can go dormant during long droughts.

Dimetrodon orients its body so that the large sail on its back faces the Sun to quickly increase its body temperature, a possible use of its elongated vertebral spines.

↖ **The compound eye of a trilobite** is made up of numerous tiny lenses of calcium carbonate, an adaptation that allowed some species to enjoy vision that spans a viewing range of almost 360°. Development of sight was an important milestone in the Cambrian. Together with the ability of an organism to make hard body parts, such as jaws and teeth, this opened the way for predatory behavior. This, of course, was followed by evolution of numerous counter-strategies, such as camouflage and armor.

← **The Atlantic puffin**, *Fratercula arctica*, has developed an extremely large beak lined with a series of inward-pointing spikes. This enables the puffin to load its beak with numerous fishes oriented crosswise. The puffin uses its rasped tongue to hold the fishes against its beak spikes while it adds successively more fishes. At certain times of the year, its beak becomes brightly colored for courtship displays. The ability of birds to evolve different shaped beaks relatively rapidly has enabled them to adapt to life in a variety of habitats.

Relating and mating

Complex relationships between and within species have been evolving ever since life began. Although sparse, evidence of relationships has been found throughout the fossil record. For many species, communication is essential for sexual reproduction to take place—methods include sporting colored crests and plumage, behavioral displays and attraction by sound or scent. For many animals, such as honeybees, communication is a vital part of the search for food. It may also be used as a form of protection—grassland squirrels use different whistling signals to alert the community if a predator is near, and when the danger has passed. Altruism is an extremely rare characteristic in the animal kingdom—vampire bats, for instance, will adopt orphans and share food outside of kin relations. Other relationships exist between different species and particularly common are different forms of deceptive behavior.

MUTUALLY BENEFICIAL RELATIONSHIPS

Mutually beneficial symbiosis, or mutualism, is a long-term working relationship between often quite unrelated species. Well-documented examples include flowers that need insects, birds or mammals to cross-pollinate and, in return, provide food. A clownfish lives within the protective tentacles of its host anemone and, in return, provides food scraps from its prey. The Nile River crocodile will open its mouth so the Egyptian plover can enter to feed on any leeches attached to its gums. Other relationships occur that are not beneficial to one of the parties (comensalism), or even detrimental, such as in the case of parasites.

Fossil evidence gives clues to behavioral characteristics. Paleontologists surmise that past behaviors were not too different from those seen today. This fossil stingray carrying two fetuses shows it gave birth to live young. *Parasaurolophus* had a tall, hollow crest that could be trumpeted in defense or for communication. The large, horned frill of *Styracosaurus* was used in courtship displays or as a visual deterrent to predators.

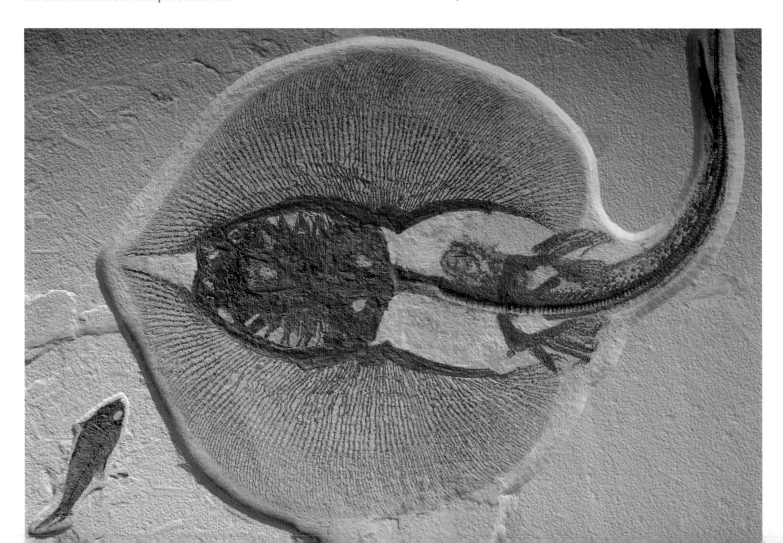

↑ **Symbiotic relationships** are common between different species. Here a red-billed oxpecker is cleaning parasites, such as ticks, fleas and flies, off the face of a willing impala. The same relationship can be found in the marine environment, where cleaner wrasse remove parasites and sometimes enter the gills and mouth of larger, otherwise carnivorous, fish.

← **Luring by deception** differs from symbiosis in that it benefits only one of the parties involved. The fly orchid flower, *Ophrys insectifera,* resembles an insect and its scent also imitates that of the female digger wasp. Male wasps trying to mate with the flower assist in the plant's pollination.

Predators

Predators take advantage of the highly concentrated protein food source gained by eating other animals. The "arms race" started before the Cambrian, with the development of hard body parts. Teeth and grasping appendages evolved side-by-side with spines and exoskeletons for protection. Highlights include *Anomalocaris*, king of Cambrian seas, through to *Tyrannosaurus rex*, top of the food web in the Cretaceous. Carnivores evolved a number of common predatory strategies to outwit and attack their prey. They use such tactics as speed and agility; forward-facing eyes for estimating distance; acute senses of sight, smell and hearing; claws, powerful jaws and pointed knifelike teeth for grabbing, cutting or tearing; and a simple, lightweight digestive system. Most important was the evolution of a larger, more intelligent brain to overcome the evolving defensive strategies of their prey.

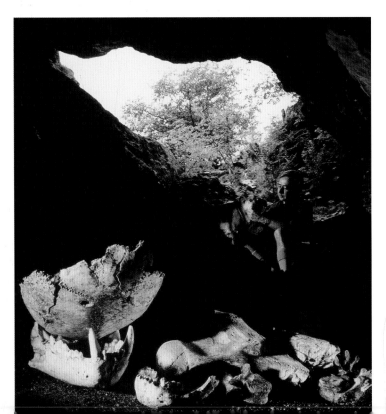

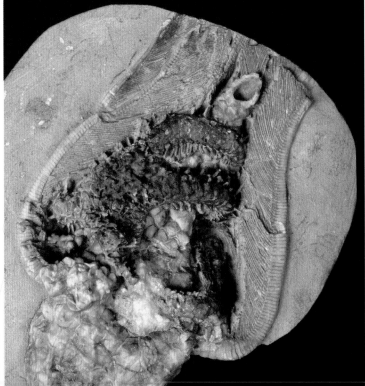

GROUP HUNTING

Group hunting is a strategy that allows predators to attack prey species that are often much larger than themselves. Small, carnivorous dinosaurs, such as *Coelophysis*, are believed to have engaged in this behavior, now used by wolves and the big cats. Prey species tend to migrate in large, social groups. They have developed forms of communication as a counter to such predatory strategies. An interesting type of group hunting is used by pods of feeding humpback whales. They circle beneath a school of tiny fishes, creating a curtain of air bubbles to corral the fishes, thereby allowing the pod to move in and feed easily.

↑   **The gray wolf,** *Canis lupus,* lives in social groups of between four and more than 30 individuals. They are very intelligent, fast, strategic pack hunters. Wolves have evolved excellent night vision, keen senses of smell and hearing, and powerful jaws.

→   **The owl** has a number of specialized adaptions that make it a highly successful predator—forward-facing eyes for stereo vision, large, sensitive eyes for night vision, excellent hearing, strong talons and feathers adapted for silent flight.

←   **Sudden burial and fossilization** have preserved the dying moments of a crinoid, *Actinocrinites gibsoni* and a starfish, *Onychaster flexilis.* The starfish is in the crinoid's calyx or cup, with its arms entwined around the anal tube of the crinoid. This 345–325 million-year-old fossil is from Crawfordsville, Indiana, USA.

←←   **Teeth marks of a leopard** in a hominid skull cap from Swartkrans cave, South Africa, show that humans were not only the predator but sometimes also the prey.

Avoiding predation

Techniques of avoiding predation evolve hand in hand with the developing hunting skills of the predator—it is this constant battle for survival that drives evolution forward. Many protective techniques have evolved. Defensive armor includes sharp spines, horns, club tails, exoskeletons and tough, armor-plated skin. Aggressive behavioral displays may be used, such as raised frills that make the animal appear larger or more dangerous. Warning colors such as the vivid blues and reds of poisonous frogs alert predators. Frightening signals and sounds, such as hissing, trumpeting, bellowing or a rattlesnake's rattle, repel intruders. Repugnant smells can deter predators, as used by skunks, lemurs and stinkbugs. Some of the more novel techniques include the ink cloud used as a smoke screen by some cephalopods, lizards that can drop a wriggling tail if attacked and then grow a new one, and the bêche-de-mer that can throw out its insides when threatened. There are also numerous cases of prey species adopting the colorings or markings of dangerous, poisonous or bad-tasting species in order to avoid predation.

COLOR CAMOUFLAGE

Many species have evolved colors such as browns, grays and greens (or white for snow) and patterns such as stripes or spots that closely resemble those of their environment. The better a species is at blending into its background, the less likely it is to be noticed and eaten. Conversely, a well-camouflaged predator can move in very close and attack unsuspecting prey, and hence feed itself more easily. The mimic octopus can change its color, pattern and shape.

The orchid mantis, *Hymenopus coronatus*, is so well camouflaged by its color and shape that it seems to be almost invisible on these pink flowers.

↑ **The panther chameleon,** *Furcifer pardalis*, uses color change for courtship displays, to show aggression, to adjust body temperature and for camouflage.

← **The Silurian trilobite,** *Phacops circumspectans*, could defend itself by rolling into a ball when threatened, ensuring that only its shell was exposed to the predator.

Venomous Maya coral snake

Non-venomous false coral snake

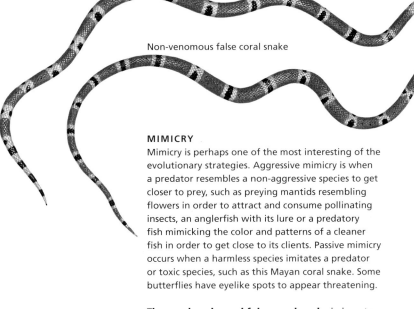

COLOR AS A WARNING

Unlike camouflage colors, which are used to hide, bright colors are usually a warning to predators to stay away. Many poisonous species use colors or patterns as a defense strategy. Examples include some frogs, fishes, snakes and fungi.

The blue poison dart frog, *Dendrobates azureus*, has few predators. A snake will writhe in agony after getting some of the skin toxins in its mouth.

MIMICRY

Mimicry is perhaps one of the most interesting of the evolutionary strategies. Aggressive mimicry is when a predator resembles a non-aggressive species to get closer to prey, such as preying mantids resembling flowers in order to attract and consume pollinating insects, an anglerfish with its lure or a predatory fish mimicking the color and patterns of a cleaner fish in order to get close to its clients. Passive mimicry occurs when a harmless species imitates a predator or toxic species, such as this Mayan coral snake. Some butterflies have eyelike spots to appear threatening.

The coral snake and false coral snake belong to completely different families of the order Serpentes. Nevertheless it has proven to be advantageous to its survival that the false coral snake looks dangerous.

Extinction

Extinction means the end of the evolutionary line for a species, family or larger group. In geological terms it is the last occurrence of an identifiable fossil. It often occurs as species transition—a natural change over successive generations into a new form. It may also occur as the sudden ending of a line. There have also been a number of major mass extinctions or biodiversity crashes where many lines end simultaneously and abruptly. In the Permian, over 90 percent of life just disappeared, opening the way for survivors to recolonize and diversify. The reason may have been a combination of massive volcanic eruptions, meteorite impacts, global climatic change and sea-level change.

HUMAN IMPACT—THE NEXT EXTINCTION

The impact of human habitation, both directly and indirectly, plays an increasingly significant part in biodiversity loss and extinction. Direct human impacts probably resulted from the overhunting of game animals such as the woolly mammoth and giant marsupials, all driven to extinction. The whales and large fishes are following close behind because of overharvesting and human disruption of the marine food web. The greatest impact, however, comes indirectly from habitat destruction caused by the burgeoning human population and the increase in post-industrial air pollution. Alarming species loss is occurring as a result of destruction of tropical rain forests for timber and farming, and the dieback of coral reefs due to global warming. The present loss rate is estimated at around 27,000 species per year and some scientists believe that Earth is currently experiencing the early stages of the next major extinction event.

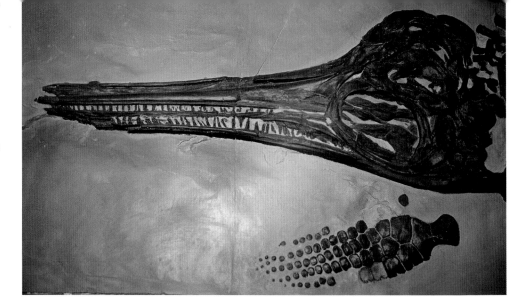

↗ **This Jurassic ichthyosaur** from Germany is one of a family of marine reptiles that, having evolved to life in the ocean, became extinct for unknown reasons during the Cretaceous.

→ **The dodo**, *Raphus cucullatus*, from Mauritius became extinct in 1670 as a result of human impact. This flightless bird was hunted by sailors for food and failed to compete with introduced species.

VULNERABILITY TO MASS EXTINCTION

What allows a certain species to survive a mass extinction is an interesting question. Statistics show that hardest hit are tropical species and marine lifeforms—coral reefs suffer extensively in every mass extinction. Plants are more resilient. It appears that highly specialized organisms, with limited environmental or dietary tolerances or those part of a fragile food chain, are the most likely to die out.

THINGS LONG EXTINCT

Life on Earth is one huge evolutionary experiment with biodiversity ultimately increasing. Nevertheless, fewer than one-tenth of one percent of all species that have ever existed are still living. The dinosaurs, such as *Phuwiangosaurus sirindhornae* (*below*) and the heavily armored *Euoplocephalus* (*right*) became extinct, along with 85 percent of all life at the end of the Cretaceous.

Lifeforms and Earth's forces

Plate tectonics and the development of life on Earth have been inexorably linked since the Precambrian. The underlying reason is that similar species will evolve in different ways when isolated from one another, eventually becoming different species. Conversely, when brought together, different species are forced to compete for the same habitat and food sources, and only the "fittest" ones will survive. The fossil record associated with the break-up of continents, such as Pangea, and the joining of others, such as India and Asia, supports this. In general, biodiversity is greater when Earth's continents are separated by ocean, much as they are today, and less when there is only one continental mass with no barriers. If tectonics push a landmass into polar regions, a continent-wide extinction will be induced, such as occurred in Antarctica.

→ **The endangered ring-tailed lemur** is found only on the island of Madagascar. It shares a common ancestry with the primates of Africa but has evolved in isolation for about 50 million years since Madagascar separated from Africa.

WALLACE'S LINE

While collecting wildlife specimens from the islands of Southeast Asia during the 1850s, Alfred Russel Wallace noted a striking difference in the fauna between the Indonesian islands of Lombok and Bali. The birds in Bali were related to those in New Guinea and Australia, while those in nearby Lombok were related to those in Java, Sumatra, Borneo and Malaya. Wallace had discovered the dividing line between the great Oriental and Australian biogeographic regions, which was later named Wallace's Line in his honor. But he remained puzzled as to why the change should occur between those two islands only a short distance apart. A century later, his line would be shown to coincide with the edge between the Indo-Australian and the Eurasian plates. Deep-sea floor at this boundary means that there was never a land bridge allowing fauna to move freely between the islands at times of low sea level. The placental mammals of Southeast Asia will have to wait another few million years for the opportunity to move into Australia and compete with the primitive mammals presently living there in isolation.

↘ **Identical _Mesosaurus_ fossils** found in similar rocks in both South America and southern Africa provide the most convincing fossil faunal evidence that the two continents were joined. This specimen is from the Karoo, South Africa.

↓ **Australia's unique wildlife** has resulted from its isolation since the break-up of Gondwana, which occurred before the arrival of placental mammals. The kangaroo is a marsupial, a primitive mammal that nurtures its undeveloped young in an external pouch.

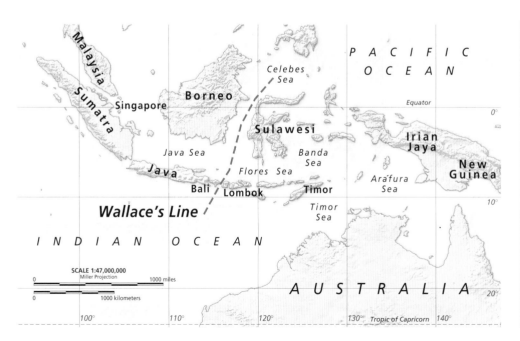

Wallace's Line

SCALE 1:47,000,000
Miller Projection

The divided continents

The break-up of the giant, single supercontinent Pangea into Earth's present continental configuration occurred during the last 180 million years. It is, however, only the final and most recent stage of a repeated series of continental break-up and reassembly. Scientists can look back dimly about 1100 million years to a time when the supercontinent Rodinia formed. It broke apart about 750 million years ago. Another supercontinent, Pannotia, formed about 600 million years ago and broke up 50 million years later. The pieces came together again 275 million years ago to form Pangea. As Pangea split up into two supercontinents, life began to evolve separately into Laurasian and Gondwanan species and continued evolving as these two landmasses further broke up. Today, six large biogeographical zones can be discerned: Nearctic (North America), Palearctic (Old World), Oriental, Australian, Neotropical (South America) and Ethiopian (south Saharan Africa). These roughly correspond to Earth's major tectonic plates.

These two widely separated landscapes—Great Smoky Mountains National Park, Tennessee, USA (*right*), and Spitzbergen, Norway (*below*)—were once part of a single geological structure—the Appalachian collision mountain belt ran across the Pangean supercontinent. About 180 million years ago, Pangea began to break apart, first in two, and then into smaller landmasses. As they moved apart, each new continent took its now-isolated flora and fauna in different directions and on different evolutionary journeys.

THE BREAK-UP OF PANGEA

The globes at right record the break-up of the supercontinent Pangea. It first separated into Laurasia (North America, Europe and Asia) and Gondwana (South America, Africa, Australia, Antarctica and India). These two landmasses split further into the continents that exist today. As Gondwana split apart, South America was pushed back into contact with North America; northern Africa was pushed into Europe; and India was rammed into Asia. Only Australia and Antarctica remain isolated.

200 million years ago

30 million years ago

Present

EVIDENCE IN THE FLORA AND FAUNA

The fossil record for the Gondwanan countries is identical until the Cretaceous. For example, the leaves of the 260 million-year-old tree *Glossopteris* (*below*), from Australia, are also found in Africa, South America, India and Antarctica. Pre-dinosaurian reptiles, such as *Lystrosaurus*, *Cynognathus*, and *Mesosaurus*, tell the same story of continents once joined. Following break-up, the fossil record in each continent became increasingly different. The Proteaceae family of plants, for instance, diversified into numerous species.

↑ ***Protea,*** from southern Africa, is a member of the Proteaceae family, which is found on all the Gondwanan continents. Plants such as *Telopea*, Australian waratahs, evolved since the break-up.

Ancient worlds

Geological time is documented in the fossil record. Periodic global disasters determined the fortunes and opportunities of life. As one massive food chain collapsed, another arose to begin anew. On a journey back to the start of time on Earth, each period of history is examined and explored.

Geological time

The passage of geological time is imperceptible to humans, who cannot even observe the minute hand of a clock moving, let alone consider rocks and mountains as anything other than solid and permanent. Minutes of the geological clock are measured in units of millions of years, each being 10,000 human 100-year lifespans. Geological time stretches back from the present to the formation of planet Earth some 4600 million years ago. It did not take the molten ball too long to cool, with zircon crystals from the primitive crust yielding dates of 4400 million years. All of this early crust has been remelted or eroded since then. Earth's oldest rocks are 3900 million year old gneisses, found in ancient continental centers, known as shield areas, that have been through numerous cycles of deformation in that time. Among the oldest fossils found, primitive stromatolite colonies date back 3500 million years in Western Australia, where living colonies can still be marveled at. Life began in earnest with an explosion in species numbers 543 million years ago at the beginning of the Cambrian period.

↑ **Earth's oldest rocks** have been found in the Isua region of Greenland. These metamorphic gneisses are an amazing 3900 to 3800 million years old.

↙ **Rock ages** are determined by measuring how much of their trace radioactive elements, such as uranium or potassium, have had time to break down.

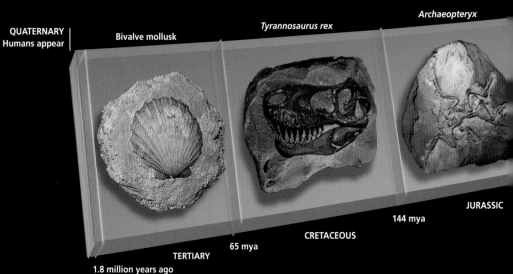

QUATERNARY
Humans appear

Bivalve mollusk

Tyrannosaurus rex

Archaeopteryx

JURASSIC
144 mya

CRETACEOUS
65 mya

TERTIARY
1.8 million years ago

RELATIVE AND ABSOLUTE TIME

Geologists place Earth's rocks onto a relative time scale depending on the rocks surrounding them. A sedimentary layer, for example, is younger than the one it overlies, but older than the one above it. If this sequence is cut by a fault or volcanic intrusion, then this activity is necessarily younger. Using fossils, an entire chronology has been built to help place rock units into their correct order the world over, even if in some places they were upside down. It was not until the development of radioactive dating techniques, however, that absolute ages of rocks were fixed in millions of years before present, thus calibrating the timeline. This great leap forward was made by Arthur Holmes in 1911, based on the principle that the unstable atoms of radioactive elements (such as uranium and potassium) found in small quantities in certain minerals (such as zircon and feldspar) randomly break down at a set rate into their "daughter" atoms (such as lead and argon). Counting the ratio of daughters to parents reveals the age.

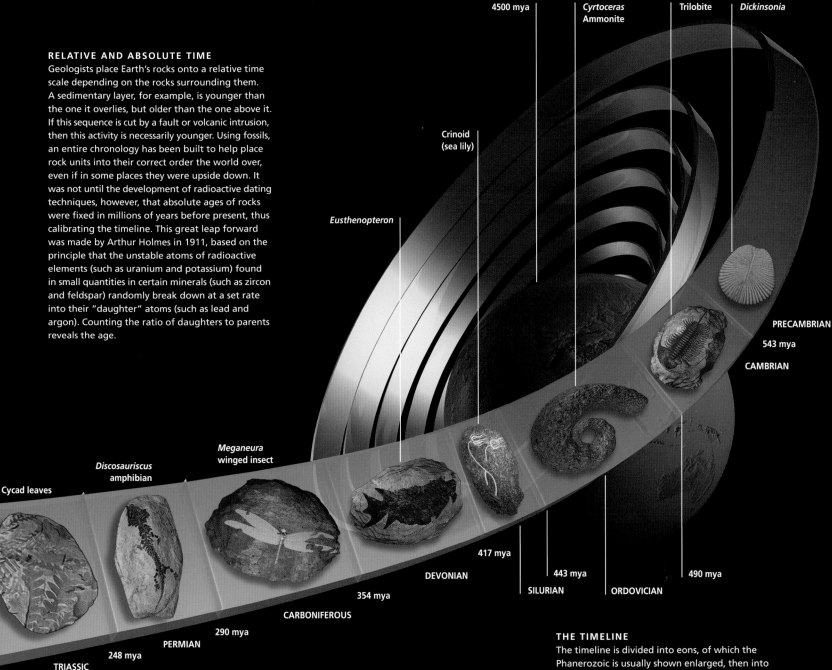

4500 mya

Cyrtoceras
Ammonite

Trilobite

Dickinsonia

Crinoid
(sea lily)

Eusthenopteron

PRECAMBRIAN

543 mya

CAMBRIAN

Meganeura
winged insect

Discosauriscus
amphibian

Cycad leaves

417 mya

443 mya

490 mya

DEVONIAN

SILURIAN

ORDOVICIAN

354 mya

CARBONIFEROUS

290 mya

PERMIAN

248 mya

TRIASSIC

206 mya

THE TIMELINE

The timeline is divided into eons, of which the Phanerozoic is usually shown enlarged, then into eras (Paleozoic, Mesozoic and Cenozoic), periods (Cambrian through to Quaternary), and epochs (not shown). The enormity of this line is realized as anatomically modern humans first appear only a pencil-line thickness from the timeline's end.

Precambrian: first life

The Precambrian is an enormous expanse of time that encompasses the first seven-eighths of Earth's entire history. This length of time is broken into three eons. The Hadean, from 4600 to 3800 million years ago, covers a time during which our Solar System coalesced into the planets and a thin crust began to form over the molten Earth. It is sometimes known as the rockless eon. The Archaean, from 3800 to 2500 million years ago, marks the beginning of the geological record, with some evidence from Greenland suggesting that first life may have appeared practically at the same time as some of the earliest rocks. Primitive cyanobacteria appeared in the seas 3500 million years ago, then quickly began to increase in number, building huge stromatolite reefs. The Proterozoic, 2500 to 543 million years ago, witnessed a rapid buildup of atmospheric oxygen levels as the result of bacterial photosynthetic activity. The evidence for this is found as seasonal layers of iron oxide in banded iron formations. High oxygen levels began to limit the growth of the very bacteria that were producing it, but made possible the evolution of the first multicellular protist life in the form of algae. The Proterozoic was dominated by a single lifeless supercontinent, Rodinia, that began to break up 750 million years ago. At the very end of the Proterozoic, during the Vendian period 650 to 543 million years ago, multicellular animal life arose.

VENDIAN OR EDIACARAN FAUNA

To date, only a handful of localities provide any insight into the evolution of the first multicellular animal life at the end of the Precambrian. The rarity of such sites is because the organisms were soft bodied and therefore not easily preserved. At Mistaken Point in Newfoundland, Canada, ash from a 565 million-year-old volcanic eruption buried an entire seafloor community. The Ediacara Hills in Australia is another important locality. Multicellular life probably arose in response to rising atmospheric oxygen levels. Seafloor forms that evolved include jellyfish and other forms so strange that they cannot be easily compared with any later fossils, or even with fossils from other Precambrian sites. Some may be related to worms or anemones but do not seem to fit into the present system for classifying life. Predatory forms had not yet developed and most animals fed on algal and bacterial mats growing on the seafloor or suspended particles. Some left tracks and burrows. They were large and flat bodied, perhaps to maximize their surface area for ingestion as they had not developed complex internal organs.

→ *Dickinsonia*, **a strange, soft-bodied** Vendian animal, was found in 570 to 543 million-year-old Precambrian rocks from the Ediacara Hills, Australia. It is one of the oldest multicellular animal lifeforms, perhaps related to the annelid worms.

↑ **During the Archaean,** about 3500 million years ago, mounds of stromatolites formed primitive reefs beneath strange anoxic skies. These calcium-carbonate-depositing photosynthetic bacteria can still be seen living in Shark Bay, Australia.

→ **During the Proterozoic,** about 1800 million years ago, oxygen buildup in the atmosphere due to bacterial photosynthetic activity allowed simple protist life, such as different forms of multicellular algae, to flourish in the oceans.

The Cambrian period (545 to 490 million years ago) saw an explosion of new marine life, with fossils becoming common in rocks dating from this time onward. Primitive calcareous sponges (archaeocyathids) built the first reefs, which were home to numerous arthropods, chiefly trilobites (of which there were more than 100 families in the Cambrian), mollusks (mainly gastropods), echinoderms, graptolites and even primitive chordates. These organisms probably evolved from soft-bodied Precambrian fauna. This rapid diversity may have occurred in as few as 10 million years. Hard body parts, such as claws, sharp mouth parts, protective exoskeletons, spikes and shells, evolved. For the first time, organisms developed eyesight. Representatives from all the animal phyla appeared during the Cambrian. Interestingly, it appears that no new phylum-level body plans have evolved since the Cambrian. During this time, global climate was warmer and more uniform than today. Extinction rates were generally high, with two major global mass extinction peaks: the first in the mid-Cambrian that depleted the trilobites; and the second, more destructive event, at the close of the period. Meanwhile, the continents remained barren.

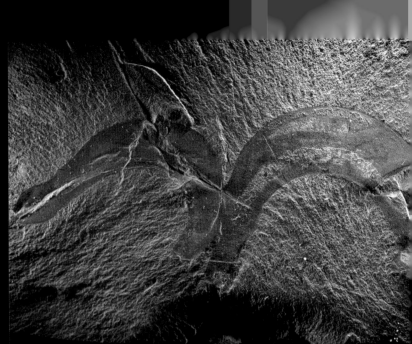

↑ ***Ottoia*, a segmented worm** from the Cambrian Burgess Shale, lived in a U-shaped burrow from which it extended its proboscis to catch prey.

↙ **Trilobites were among the first arthropods** to evolve sight, assisting their search for food. Their compound eyes had lenses of clear calcium carbonate.

→ **The oceans at the beginning of the Cambrian** were teeming with newly developing, soft-bodied animal life, some not unlike present-day jellyfish.

THE CAMBRIAN BURGESS SHALE
The Burgess Shale deposit is in Yoho National Park, British Columbia, Canada. It offers a unique glimpse into life in the Cambrian because of its excellent preservation of unusual soft-bodied animals and other arthropod-like armor-plated and spiked forms. Many of these, which rapidly became extinct, are quite alien in appearance. Approximately 505 million years ago, marine landslides on the western continental margin of Laurentia carried entire Cambrian bottom-dwelling communities from their relatively shallow location at the edge of a carbonate reef platform into deep water at the foot of the platform, where they became preserved without being disturbed. Other significant and even earlier Cambrian sites (520 to 515 million years old) have been found at Chengjiang in China and Sirius Passet in Greenland.

Cambrian mass extinction

The extinction at the end of the Cambrian severely impacted on many of the unusual lifeforms that had exploded into existence at the beginning of the period. Most affected were the trilobites, reef-building sponges and brachiopods. While the exact cause is unclear, glaciation at the time may have significantly lowered temperatures and possibly also the oxygen content in the water, killing species unable to adapt to the changed conditions. The accompanying sea-level fall would also have reduced the shallow water habitat on the continental shelves.

THE CAMBRIAN EXTINCTION SCENE

For some time, ice shelves encroaching on the Cambrian seas have caused temperatures to drop and reduced shallow-bottomed habitats. Life has become difficult and food scarce. The sponges and archeocyathids are dying back and crumbling. Armored *Halkieria* scavenges among the remains of a trilobite exoskeleton while the blind, bizarre spiked *Hallucigenia* pauses in its search for food, sensing predators nearby. Other strange bottom dwellers, such as the armor-plated, spiked *Wiwaxia*, feed off algae on the sand. Trilobites were one of the first of the Cambrian creatures to develop sight; one burrows under one of the *Wiwaxias*, trying to overturn it and attack its unprotected underbelly. A wormlike *Pikaia* undulates through the water, filter-feeding. Above, a flight of four *Marella* lace crabs dart into the path of a hungry, camouflaged *Anomalocaris* lurking in the shadows of the iceshelf. *Anomalocaris*, a segmented arthropod-like predator, was king of the Cambrian seas. It lunges out, ensnaring the leading *Marella*.

Trilobite, bottom-dwelling arthropod **G0**
Archaeocyathid, reef-forming sponge relative **J2**
Hallucigenia, 14-legged, spiny creature **C2**
Halkieria, armor-plated creature with shell **B0**
Wiwaxia, creature with protective spines **F0, J0**
Marella, primitive arthropod lace crab **G3, H3**
Pikaia, earliest representative of phylum chordata **J3**
Anomalocaris, predatory protoarthropod **G5**
Vauxia, soft branching sponge **C4**
Green algae, protist cells with chlorophyll **A2**

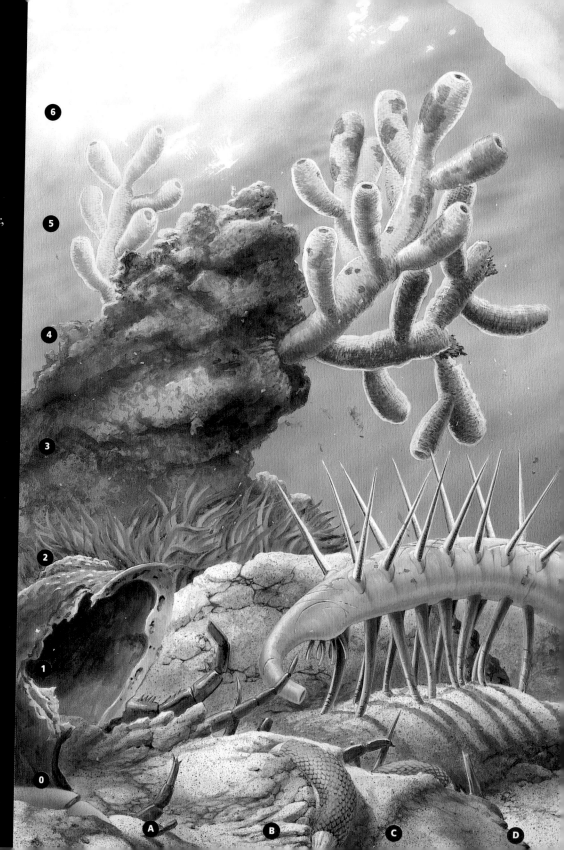

Ordovician: peak biodiversity

Following the extinction at the end of the Cambrian, the Ordovician period (490 to 443 million years ago) saw the rapid development of new invertebrate animals, which occupied the niches left by their predecessors. The rate of diversification of marine families peaked at this time. Carnivorous scorpion-like arthropods called eurypterids, some larger than a human, roamed the seas, lagoons and swamps. Some may have even spent part of their lifecycle out of the water. Rugose and tabulate corals and stromatoporoids, bivalve mollusks and planktonic graptolites increased in number and diversity. Other filter feeders, such as the crinoids and brachiopods, also developed. The intelligent nautiloid cephalopods evolved along several lines, replacing *Anomalocaris* as the top marine predator. Ordovician trilobites diversified in shape and form. The first land plants achieved a tenuous toehold in damp coastal areas of the bare continents. These were tiny, primitive, non-vascular bryophytes or mosses that evolved from near-shore green algae.

A COLD TIME IN EARTH'S HISTORY

Oceans separated the large continents of Laurasia, Baltica, Siberia and Gondwana, fragments of the original supercontinent Rodinia. The early Ordovician climate was warm enough that extensive reef complexes grew in the tropics. However, as the great Gondwanan continent drifted over the South Pole, conditions began to change. Sea levels fell, exposing continental shelf areas as huge ice sheets built up on Gondwana. Temperatures dropped to some of the lowest in Earth's history as glaciers pushed outward from polar regions. In these increasingly difficult conditions of cooler seas and restricted shallow-water shelf habitats, species—particularly those in the tropics—unable to adapt quickly enough began to suffer. These factors culminated in a mass extinction at the end of the Ordovician period.

↑ **Graptolite animals** each live in a slanted cuplike theca, which gives the colony a saw-bladelike appearance. Filter-feeding graptolite colonies float around in the sea. They arose in the Ordovician.

← **Sea scorpions,** or eurypterids, are segmented aquatic arthropods that thrived from the Ordovician to the Permian, some of them reaching 6 feet (1.8 m) in length. Some may have breathed air and water, possibly emerging onto land for part of their lifecycle. These fossils are from Silurian deposits in New York State, USA.

→ **Massive evaporite deposits** form in basins at times of low sea level such as existed during the Ordovician. These landlocked basin lakes, including Australia's Lake Eyre, became increasingly salty.

Ordovician mass extinction

More than 100 entire families perished in the mass extinction at the end of the Ordovician, about 440 million years ago. The worst affected were trilobites, nautiloids, conodonts, bryozoans, graptolites, reef-building corals and brachiopods. This extinction event is attributed to extensive glaciation caused by the continent of Gondwana passing over the South Pole. As with the end of Cambrian mass extinction, this caused a fall in sea level, a loss of shelf habitat and a drop in temperature and oxygen levels.

THE ORDOVICIAN EXTINCTION SCENE

Earth's cooling seas brace for another extinction event. The sea level is dropping as water becomes locked up in ice on land. Ecosystems are under stress. Hungry, straight-coned nautiloids move in, scattering a group of trilobites. They attempt to escape their predators by swimming, burrowing or rolling into a tight ball. The late Ordovician seafloor is covered with flexible-stemmed crinoids (sea lilies), tabulate corals and large, solitary, horn-shaped rugose corals. The brachiopods are also flourishing. Numerous graptolites float in the waters above. Some of the trilobite species have moved off the seafloor, such as *Sphaeragnostus* and the large-eyed *Opipeuter,* which swims along on its back. Icebergs float in the sea.

Cameroceras, straight-shelled nautiloid **G4**
Treptoceras, straight-shelled nautiloid **D1**
Tentaculites, often classed as a mollusk **H6, B6**
Opipeuter, large-eyed swimming trilobite **C5**
Isotelus gigas, burrowing trilobite **D2**
Sphaeragnostus, floating, rolled trilobite **G6**
Pseudosphaerexochus, swimming trilobite **F4, G5**
Sowerbyella, strophomenid brachiopod **H0**
Climacograptus, floating graptolite **E6**
Orthograptus, floating graptolite **D6**
Nemagraptus gracilis, graptolite **B4**
Dicellograptus, floating graptolite **C6**
Xenocrinus, camerate crinoid **A1**
Ectenocrinus, inadunate crinoid **G1**
Cyclonema bilix, sea-bottom gastropod **K2**
Calapoecia, colonial tabulate coral **K0**
Grewingkia, solitary rugose coral **J3**

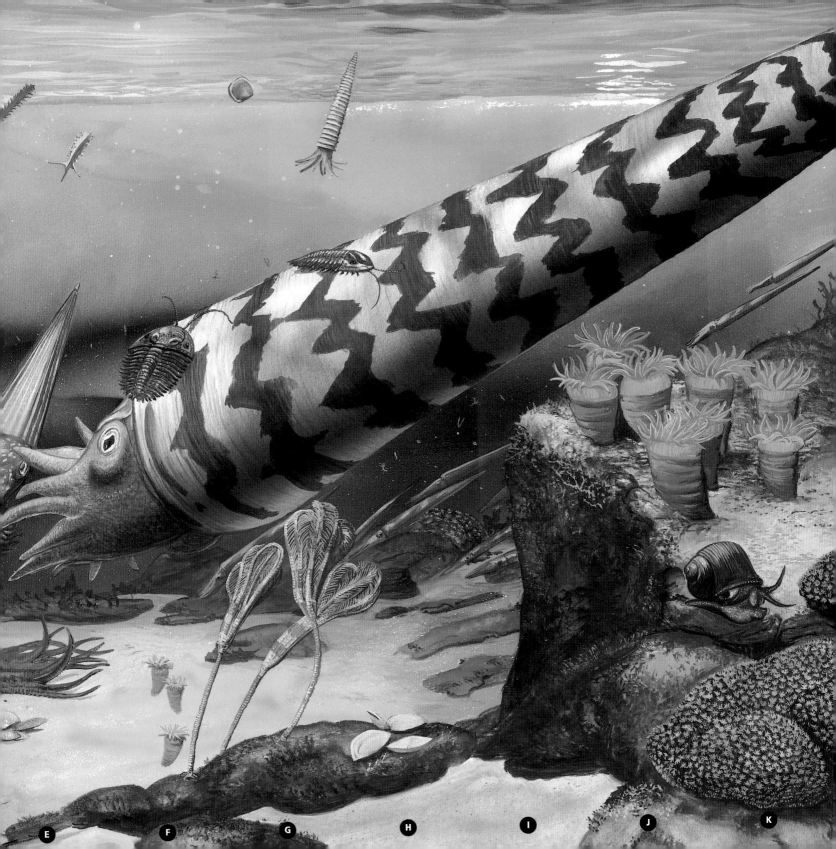

E   F   G   H   I   J   K

Silurian: plants invade the land

Primitive plants continued to colonize the land during the Silurian period (443 to 417 million years ago), with the first fungi and vascular plants—those with conductive tissue for water and food—appearing in the late Silurian. The evolution of vascular tissue paved the way for the development of true leaves, roots and stems. Rare late Silurian arthropod fossils suggest that primitive mites, spiders and centipedes were colonizing coastal land. Diversification of marine life continued after the setback caused by the extinction of many species at the end of the Ordovician. The calcareous sponges and corals built huge limestone reefs, while the nautiloids, brachiopods, bryozoans, graptolites and crinoids continued to diversify. Trilobites, however, remained reduced in number. Cartilaginous fishes, which appeared in the Ordovician, continued to diversify and evolved a lower jaw. Strange armor-plated placoderms, such as *Dunkleosteus*, also appeared and later dominated the Devonian seas. Some marine fishes evolved into and went on to dominate freshwater varieties.

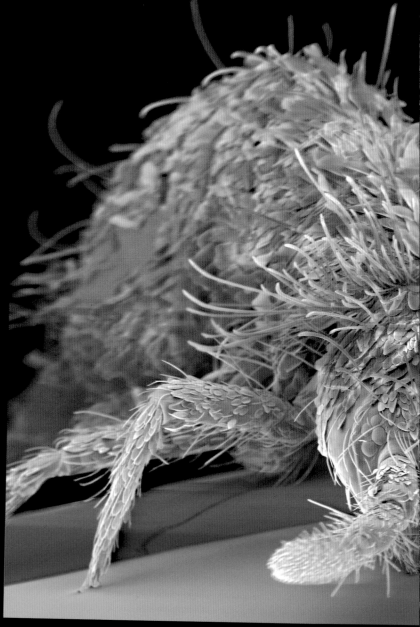

↑ **Springtail-like insects** may have been the first arthropods to colonize land. *Collembola* feed on plant litter and are considered living fossils.

← **Crinoids or sea lilies,** such as *Scyphocrinites elegans*, were attached to the sea bottom with a stem and captured food using feathery arms. Only a few hundred species still exist.

SEA LEVELS RISE AGAIN

After the late Ordovician glaciation event, warming and melting of the continental ice sheets caused sea levels to rise again. The flooding of the continental shelves created new shallow-water marine habitats and favorable conditions for coral-reef building. The continents were still separated by ocean, but had started to drift back together. About 425 million years ago, Laurentia collided with Baltica to form Euramerica. The crust between the

mountain belt under the severe compressive stress, in a fashion similar to that forming the Himalayan mountains of today. The remains of this Silurian range form the Appalachians of the USA and the corresponding (but now separated) mountain belts in present-day Greenland, Scotland, Ireland and Norway. This collision was part of a larger mountain-building event and is known as the Caledonian Orogeny, which continued for

↑ **Primitive, non-vascular plants** such as mosses, having evolved from green algae, began to colonize the land in earnest during the Silurian. These plants remained small as they did not have the structure needed to support a larger plant. They were unable to grow outside moist environments.

Devonian: the age of fishes

The Devonian period (417 to 354 million years ago) was witness to the diversification of fishes and their dominance of the seas. Most abundant were the armor-plated placoderms, primitive sharks, lungfishes and lobe-finned fishes. The bony fishes, which today are the largest and most diverse group of all the vertebrates, arose at about the start of this period. Ammonoids appeared and other invertebrates flourished, while sponges and corals built some of the largest reef complexes in the world. On land, the evolution of seeds meant that plants were no longer dependent on moisture to reproduce. Plants pushed inland from pockets of mosses and horsetails by the water's edge and went on to colonize vast areas. By the end of the Devonian, forests of *Archaeopteris*, the first tall trees, were abundant. The first amphibians, which probably descended from either lungfish or rhipidistian fish, began to move onto the land.

↓ **Skull of *Eastmanosteus***, a Devonian armored placoderm. These fishes did not have teeth, but a series of sharp, bony plates for cutting flesh. They disappeared in the Devonian mass extinction.

↓ **The Devonian Canning Basin reef**, Australia, is one of the largest limestone reef complexes. It was built by stromatoporoids and other calcium-carbonate-producing micro-organisms. The outcrop extends over 200 miles (320 km).

MILD CLIMATE AND HIGH SEA LEVELS

Warm and mild conditions worldwide helped the dramatic colonization of land by plants. Soil formation was accelerated, and the terrestrial invertebrates evolved accordingly. Sea levels were high, covering the continental shelves with shallow, warm water, thereby creating ideal conditions for the diversification of marine, freshwater and estuarine life. There was much reef-building activity. Three major continents existed at this time: Euramerica (sometimes called the Old Red continent or Laurussia), Gondwana and Siberia. They were surrounded by subduction zones and slowly drifted together. It is possible that glaciation at the end of the Devonian may have precipitated the next mass extinction.

The Devonian oceans were home to sharks and other cartilaginous fishes. Sharks are highly tuned to their environment, with acutely developed predatory senses of sight, hearing, smell and the ability to detect tiny vibrations and weak electrical fields. The hammerhead is a relatively modern shark.

Devonian mass extinction

The Devonian mass extinction crisis hit the marine community significantly some 365 million years ago, but did not much alter the terrestrial scene. The lifeforms most affected were the reef-forming rugose and tabulate corals plus the stromatoporoids. So severe were the effects that further major reef-building did not recommence until the evolution of modern corals at the start of the Permian. Completely lost were the armored placoderm fishes, while jawless fishes, brachiopods, trilobites, ammonites and conodonts were severely affected. While the cause is uncertain, there is evidence of glaciation in Gondwana; a meteorite impact has also been implicated.

THE DEVONIAN EXTINCTION SCENE

Dunkleosteus, the heavily armored placoderm and the first of the jawed fishes, was king of the seas until the end of the Devonian. *Dunkleosteus* did not have teeth but possessed a highly effective series of razor-sharp bony plates. Here, it has just made a kill and is being harassed by a school of frenzied primitive sharks sensing blood in the water. The male *Stethacanthus* is clearly distinguishable by its bizarre, flat-topped dorsal fin and head, both covered with tooth-shaped scales. These sharks would normally steer well clear of *Dunkleosteus* for fear of becoming its prey—they came to dominate the seas only after this extinction. On the seafloor, a group of *Huntonia* trilobites forage for food among the brachiopods, sponges, tabulate and rugose corals of the slowly dying Devonian reef.

Dunkleosteus, armored placoderm, jawed fish **B5**
Ctenacanthus, shark with twin dorsal spines **B2**
Stethacanthus, primitive hybodont shark (male) **E5**
Cladoselache, primitive shark, three-pointed teeth **I5**
Huntonia, trilobites **G1, J0, K2**
Tabulata, reef-forming colonial corals **I3**
Solitary rugosa, large, tubular horn corals **D1, H4**
Solitary rugosa, large, single horn coral (dead) **K0**
Colonial rugosa, reef-forming corals **H1, H0**
Colonial rugosa, corals with tentacles extended **I2**
Pentamerida, hinged brachiopods **F0**

Carboniferous: the age of swamps

The Carboniferous (354 to 290 million years ago) is commonly divided into two sub-periods: the lower Carboniferous, or Mississippian; and the upper Carboniferous, or Pennsylvanian. After the Devonian mass extinction, life again flourished in the oceans. Highly coiled nautiloids replaced the increasingly rare, straight-shelled forms; complex ammonoids replaced the simple Devonian forms; and an amazing variety of sharks and bony fishes replaced the armored and lobe-finned fishes. Life on land also expanded in the warm conditions of the Carboniferous. Spore-bearing club mosses, tree ferns and horsetails, and seed-producing plants, including seed ferns, dominated the extensive coal-producing swamps for which this period is most famous. Later, tall conifers evolved and began to dominate the forests. Amphibians thrived in the swamps and estuaries, while giant dragonflies, millipedes and other insects lived in the forest and among the organic litter on forest floors. In the middle of the period, reptiles evolved from the amphibians. Their hard-shelled eggs and protective, scaly skin enabled them to travel farther from water and colonize the land.

FORMATION OF THE SUPERCONTINENT PANGEA

The early part of the Carboniferous was mostly warm, with the abundant forests and swamps extracting vast amounts of carbon dioxide from the atmosphere and locking it away as carbon in the coal measures. Oxygen levels were higher in the Carboniferous than at any other time in Earth's history. Later, as Gondwana again drifted toward the South Pole, the formation of vast ice sheets triggered a pronounced cooling, particularly of the polar regions. During the Carboniferous, the continents of Euramerica and Gondwana continued their convergence and collision, which marked the beginning of the Pangean supercontinent. This single landmass would dominate the planet's climate and biodiversity through to the Cretaceous period.

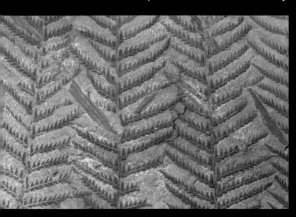

↑ **Primitive ferns,** such as *Pecopteris*, were widespread in swamps during the Carboniferous. Many of these varieties are now extinct.

→ **Reptiles evolved** from amphibians in this period. The primitive New Zealand reptile, the tuatara, is considered to be a living fossil.

→→ **Widespread swamps** in the Carboniferous laid down this period's massive coal deposits.

Permian: end of an era

During the Permian period (290 to 248 million years ago), life seemingly continued as it had through the Carboniferous. However, there was a slight decline in biodiversity levels because of decreasing opportunities for divergent evolution resulting from the coming together of the Pangean supercontinent. On land, drier conditions that prevailed toward the middle of the period saw the end of the extensive swamplands. Gymnosperms, or seed-bearing trees, such as ginkgos and conifers, became predominant. In the southern part of Pangea, *Glossopteris* flora evolved. Evaporite minerals began to accumulate in desert basins in the dry western part of Pangea, and the increasing aridity also affected vertebrate evolution. Amphibian numbers declined as the swamps dried, and reptiles took over as the dominant land animals. Pelycosaurs evolved, such as the huge sail-backed *Dimetrodon* and *Edaphosaurus*. The mammal-like therapsids also began to evolve. They started competing with and displacing the less-efficient pelycosaurs. Some may even have had fur and warm blood, making them far more adaptable to cold conditions than their more primitive forebears. In the sea, coral reefs continued to proliferate and support marine invertebrate communities; sharks remained as the dominant predators. Ammonoids developed increasingly complex shells, while the decline of the trilobites continued.

THE PANGEAN SUPERCONTINENT

The massive continental assembly of Pangea was completed when Siberia collided into northern Europe. This pushed up the Ural mountain belt. Pangea was a landmass shaped like a huge letter C, partially enclosing a body of water called the Tethys Sea and surrounded by a single ocean known as Panthalassa. Away from the coast and mountains, the vast interior of Pangea was probably dry, with great daily and seasonal temperature variations because of the lack of the moderating effects of the sea. Glaciers retreated and the climate became warmer and drier toward the later part of this period.

→ **The Permian–Triassic extinction** boundary is visible in the wall of Butterloch Canyon, Italy, as an erosion notch associated with a band of light sediment and a change in bedding thickness.

↓ *Dinogorgon rubidgei* is one of the many therapsids (mammal-like reptiles) that did not survive the Permian mass extinction, the most disastrous extinction known.

Permian mass extinction

In this, the gravest of Earth's mass extinctions, 90 percent of marine life and some three-quarters of vertebrate life on land was wiped out. Fossil-rich layers mark the end of the Permian around the world. Coral reefs were obliterated, taking another 10 million years to recover. A long list of lifeforms were lost forever, including the trilobites, tabulate corals, rugose corals and blastoids. Even more were decimated—sharks, eurypterids, bony fishes, crinoids, bryozoans, brachiopods and ammonoids. Life took 150 million years to regain the diversity it had in the Permian. Volcanic eruptions in Siberia, with associated greenhouse-gas global warming, were possibly responsible for this extinction.

THE PERMIAN EXTINCTION SCENE

Time is running out for nearly all late Permian life. The sky is an eerie purplish red as sunlight weakly filters through an atmosphere laden with volcanic dust and greenhouse gases. On land, a pair of saber-toothed mammal-like therapsids attack three bony-plated scutosaurs, which are themselves weakened and unable to find food in the ruined araucarian forest. In the sea a huge whorl-toothed shark, *Helicoprion*, has moved in from the ocean in search of food and is harassing a sharklike ray, which takes flight through a meadow of long-stemmed sea lilies. Three ray-finned bony fishes move in to see what has been stirred up. On the bottom, the last of the trilobites, *Phillipsia*, hunts among the brachiopods.

Helicoprion, whorl-toothed shark **E2**
Flexibilia, an order of stemmed crinoid, sea lily **A1**
Camerata, an order of stemmed crinoid, sea lily **B1**
Menaspis armata, sharklike chimera **B2**
Palaeoniscum, ray-finned bony fish **A4, B4, C4**
Phillipsia, last trilobite **A0, E0**
Hebertella, orthid brachiopod **B0**
Rhynchonellida, brachiopod **A0**
Composita, spiriferid brachiopod **A0**
Tabulate coral **B0**
Inostrancevia, carnivorous therapsids **G4, J4**
Scutosaurus, bony-plated anapsid reptiles **J3, K3**
Araucaria, various species of conifers **I6**

E   F   G   H   I   J   K

Triassic: reptiles rise to dominance

The Triassic period (248 to 206 million years ago) followed the almost-total Permian extinction and marked the start of a new era, the Mesozoic. Life had to begin almost anew, both in the sea and on the land. The massive calcareous sponge reefs, and tabulate and rugose corals had disappeared, making way for modern scleractinian ▯s. It was a slow process of rebuilding. Ammonoids, brachiopods, bivalves and echinoderms recovered, and squidlike belemnites appeared. Many reptiles, such as the nothosaurs and ichthyosaurs, returned to the seas as active predators. On land, the conditions remained dry. Pangea's interior was covered by stunted drought-tolerant vegetation and dry salt lakes. Coastal forests were dominated by cycads and conifers. During the Triassic, reptiles such as the cold-blooded archosaurs dominated the warm-blooded therapsids, perhaps due to their better adaptability to the arid conditions. The first turtles, dinosaurs and crocodiles arose. Reptiles also took to the skies, with some of the pterosaur species reaching gigantic proportions. At the end of the period, true mammals appeared. They were small rodent-like, nocturnal creatures that remained insignificant in comparison to the reptiles until the end of the Mesozoic era.

↑ **Paleontologists** at the Sahat Sakan Institute at Kalasin in Thailand study the bones of a young sauropod dinosaur from the Triassic.

↑ **Turtles** appeared during the Triassic, and later evolved features such as paddle limbs and a lighter shell, enabling them to adapt to life in the sea. Turtles still come out of the water to lay eggs.

PANGEAN DESERTS AND EVAPORITES

Ice was disappearing from the poles and the climatic trend of warming and drying continued from the Permian, particularly in Pangea's vast interior. The rate of accumulation of evaporites, such as salt beds, was greater than at any other time in Earth's history. Although new species evolved after the catastrophic Permian extinction, biodiversity remained at an all-time low. A series of minor extinctions seems to have taken place as species, previously living in diverse continents, continued to come into contact and compete with one another on the Pangean supercontinent. Uplift and rift-valley faulting between the northern Laurasian and the southern Gondwanan sections heralded the first rumblings of the break-up of Pangea that continues today.

← **Evaporite deposition** reached its peak during the hot, dry conditions of the Triassic. Lake Amadeus in central Australia is the world's largest, internally draining, ancient evaporite basin. Here, white expanses of salt surround slightly higher islands of stunted, drought-tolerant vegetation.

Triassic mass extinction

Another smaller setback for the evolving amphibians and reptiles occurred at the end of the Triassic, 206 million years ago. Though only a quarter of all terrestrial and marine life was affected, it heavily impacted on the large amphibians, most of the mammal-like reptiles (therapsids) and the non-dinosaurian reptiles. It is thought that this event may have cleared the path for the evolution of the dinosaurs to take up their position of dominance in the Jurassic and Cretaceous periods. It is not clear if the loss of species occurred as a single event or a number of consecutive, smaller events. The reasons for this extinction are not known, but significant climate changes, such as an increase in rainfall, are often cited.

THE TRIASSIC EXTINCTION SCENE

Two top Triassic predator species hunt for food along some river flats. *Postosuchus* has spotted a group of aquatic reptiles despite their camouflage, and they scatter into the water for protection. On the other side of the stream, an *Ornithosuchus* has caught a lizard; its partner rears back at the sight of *Postosuchus*. Reptiles dominate the land, sea and air. An amphibian, one of the last of the giants, lies in the horsetails next to a clump of cycads. By basking in the sun, it adjusts its body temperature. A pair of *Coelophysis*, lightly built, fast-moving pack hunters, herald the emergence of the dinosaurs.

Ornithosuchus, primitive thecodont archosaur **B3**
Postosuchus, armored carnivorous reptile **K4**
Paracyclotosaurus, giant amphibian **B1**
Equisetales, primitive horsetails **D1**
Lariosaurus, nothosaur, an aquatic reptile **F1**
Placerias, tusked mammal-like reptile **F2, G2, H2**
Coelophysis, early theropod dinosaur **G1, I1**
Desmatosuchus, horned herbivorous reptile **F3**
Kuehneosaurus, gliding lizard **J6**
Eudimorphodon, pterosaur, flying reptile **F6**
Palaeocycas, primitive cycad **D0, K0**
Glossopteris, giant tree fern **F4**
Araucarioxylon arizonicum, conifer tree **I5**
Ginkgo biloba, maidenhair tree **K5**
Walchia, ancient conifer tree **B5, C5**

E   F   G   H   I   J   K

Jurassic: rise of the dinosaurs

The Jurassic (206 to 144 million years ago) was a period of growth in species numbers after the Triassic. Reefs of scleractinian corals grew abundantly in the warm tropical seas, with sponges, bryozoans, gastropods, bivalves, ammonoids and belemnites. Brachiopods and crinoids continued, but in much-reduced numbers. Bony fishes were plentiful, along with sharks and rays. New marine reptiles—ichthyosaurs, plesiosaurs and pliosaurs—developed from their Triassic predecessors. On land, gymnosperms and cycads continued to thrive, and numerous groups of herbivorous insects evolved. Dinosaurs were diverse and abundant, some being among the largest animals that have ever lived. A variety of meat-eating dinosaurs preyed on the plant-eaters. Some may have had warm blood and insulating feathers. Later, as is aptly demonstrated by *Archaeopteryx*, these feathers evolved for flight. Early mammals remained in the background, but continued to develop and diversify.

↓ **This coiled ammonoid cephalopod** was found in late Jurassic strata of the Saratov region, Russia. Ammonoids and belemnites were the dominant invertebrates in the seas. The shell of this specimen still shows its original luster and sheen.

↑ **At the Palisades** of New Jersey, USA, extensive basaltic volcanism occurred along the rift zone as Pangea split into Laurasia and Gondwana, creating lava flows and extenssive dikes and sills, here exposed by the Hudson River.

↗ ***Gasosaurus constructus*** was a carnivorous Jurassic dinosaur about 11½ feet (3.5 m) long. It probably walked on its two hind legs, with the weight of its forward-leaning body balanced by its heavy tail.

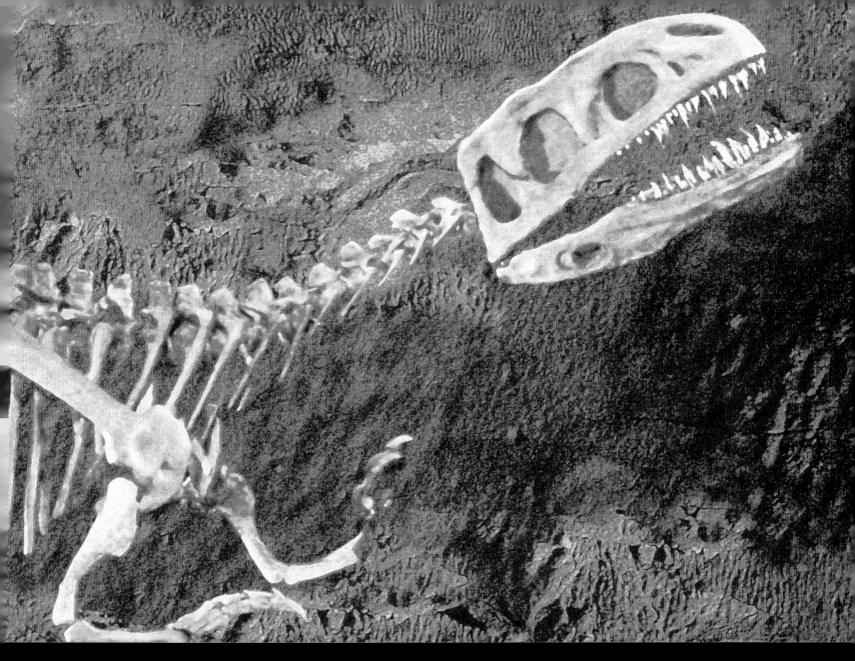

PANGEA SPLITS: WARM TROPICAL CONDITIONS PREVAIL

The rifting of Pangea that began in the late Triassic continued into the Jurassic, with a long, shallow seaway appearing between Laurasia and Gondwana. First the ancestral Gulf of Mexico opened, then the proto-North Atlantic ocean. In Gondwana, rifting began forming narrow seaways between South America, Africa and Antarctica. Ancestral India began to break away from Africa. These rifts and seaways looked a lot like those of today's Great African Rift, Gulf of Aden and the Red Sea. Volcanoes associated with the Jurassic rifts erupted huge quantities of basalt. The ice caps melted from the poles, the climate was warm and the sea level was high. Shallow seas spread across Europe and Russia, and much of central No th A series was flooded. This a d the numerous new

seaways, meant that seasonal climatic variations were reduced, the large deserts disappeared and conditions on land were moist. Warm tropical seas saw an explosion of life at the base of the food chain and a consequent rise in marine biodiversity. Vast numbers of plankton sank to the bottom and formed the large petroleum deposits of the Gulf of Mexico and the North Sea. Separation of the new continental fragments by sea, and favorable climate on land, meant that the rate of terrestrial biodiversity increased markedly. This increase has continued to the present day, and is only recently being reversed by the combined impacts of human population growth and massive growth in industrialization.

Cretaceous: flowers bloom

Dinosaurs continued to flourish during the Cretaceous (144 to 65 million years ago), and new species of plant eaters and carnivores evolved, including a large number of dinosaur-like protobirds and birds. The crested dinosaurs (ceratopsians) appeared, as did the club-tailed ankylosaurs and the tyrannosaurs. Birds appear to have pushed the pterosaurs to extinction. Mammals, including monotremes, marsupials and placentals, were on the rise. In the water, mosasaurs and turtles were common, modern crocodilians had appeared, and plesiosaurs had developed some new forms, including the huge, long-necked *Elasmosaurus*. Ichthyosaurs, however, were reduced in number and had died out by the end of the period. Amphibians consisted mainly of the existing groups of frogs and salamanders. By the end of the Cretaceous, the fast and efficient teleost fishes had become dominant in marine and freshwater habitats. One of their key adaptations was a fully movable top and bottom jaw that enabled their jaws to protrude when opening. This greatly assisted them in capturing prey.

↓ **Flowering plants** (angiosperms) increased in abundance toward the end of the Cretaceous. They gradually displaced the ferns, cycads and conifers (gymnosperms) from their long-established niches. The presence of flowering plants stimulated insect evolution and, by the end of the period, many of the modern symbiotic relationships between plants and insects had appeared. These beautiful fields of flowers can be seen along the Welsh coastline.

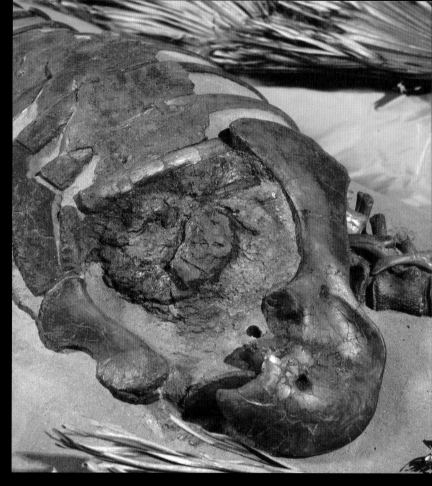

Cretaceous mass extinction

The second-largest mass extinction in geological history was also the most recent, occurring 65 million years ago. Among the 85 percent of living things that disappeared were the dinosaurs, pterosaurs, marine reptiles and ammonites. Many of the niches left empty were taken up by mammals. Trace amounts of the metal iridium found in rock strata of this age suggest two causes: either the giant meteorite that hit Earth, creating the Chicxulub crater in Mexico, or the massive volcanic eruptions of the Deccan Traps, India. Perhaps both initiated this extinction.

THE CRETACEOUS EXTINCTION SCENE

A *Tyrannosaurus* family moves in search of a meal, flanked by three small, fast, birdlike *Troodon*. These intelligent creatures know there are always leftovers after a big kill. The pack-hunting predators cause a stir wherever they go. A herd of long-crested hadrosaurs that had been grazing in the clearing stampede into the forest, trumpeting in fear. Smaller *Carnotaurus* looks up from its meal, hoping it will not be noticed, while the defensively armor-plated and horned herbivores continue browsing. Plesiosaurs and mosasaurs dominate the sea, eating fishes and other invertebrate marine life. Even the ammonite's shell is no match for their powerful jaws. Tooth-beaked flightless birds also dive for the abundant fishes.

Pteranodon, flying pterosaur **D6**
Tyrannosaurus, large carnivorous therapod **D5, E5**
Troodon, small, intelligent, birdlike therapod **D4**
Hesperornis, flightless aquatic bird **A1**
Elasmosaurus, long-necked plesiosaur **D0**
Perisphinctes, large, coiled ammonite **E1, F0**
Plotosaurus, fish-eating mosasaur **G1, H1**
Belemnites, straight-shelled cephalopod **G0, H0**
Parasaurolophus, duck-billed crested hadrosaur **G4**
Carnotaurus, carnivorous theropod **K4**
Styracosaurus, horned, frilled ceratopsian **A3, C3**
Nodosaurus, armored ankylosaur **H3**
Quetzalcoatlus, largest pterosaur **F6**
Agathis, kauri pine **B6**
Araucaria bidwillii, giant coned bunya pine **A6, G5**
Metasequoia glyptostoboides, conifer **K6**

E   F   G   H   I   J   K

Tertiary: age of mammals

The Tertiary (65 to 1.8 million years ago) marks the beginning of the Cenozoic era, a time when mammals moved into the niches left vacant by dinosaurs. The Tertiary is divided into five epochs—the Paleocene, Eocene, Oligocene, Miocene and Pliocene. The Paleocene began with dense forests growing in a warm, moist climate but, by the end of the Eocene, temperatures had cooled. Open woodlands replaced the forests. Small mammals were common and the first primates arose. Some mammals, such as bats, took to the air; others, including whales, adapted to life in the sea. In the Oligocene and Miocene, the climate became increasingly cooler and more seasonal as Antarctica froze over. Sea levels dropped and animals migrated between continents. Flowering plants adapted to the conditions and grasses became one of the most important groups. They supported diverse populations of large grazing mammals, including horses, deer and antelope. Hominids separated from the chimpanzee lines. In the Pliocene, cooling continued and the land bridge between North and South America formed. Grazing herbivores continued to grow larger and prosper, but faced fierce predators, such as saber-toothed cats, pack-hunting dogs and bears. Then, about two million years ago, a global ice age began.

↓ **This fossil horse** lived in the Eocene, about 49 million years ago. It was about the size of a fox and probably foraged in the forest for leaves and fruit. It was found in the Grube Messel open pit oil shale mine near Darmstadt, Germany.

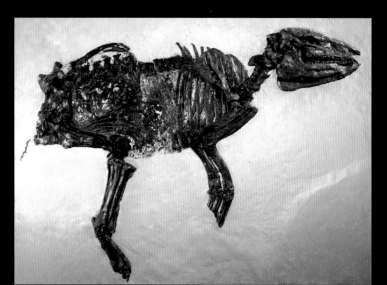

AN APPROACHING ICE AGE

During the Tertiary, the northward movement of Africa pushed the fragments of southern Europe together, thrusting up the Pyrenees, Iberian Plateau, Alps, Balkans and Atlas mountains. India collided with Asia, raising the Himalayas. Meanwhile, Australia broke away from Antarctica, leaving it isolated and over the South Pole, where winter snow did not melt in summer. The buildup of ice on Antarctica, coupled with mountain building worldwide, precipitated an ice age at the end of the period. The evolution of Tertiary life was driven by these tectonic events. Forest animals adapted to the open plains, and some gigantic mammals evolved. As glaciers grew, more water was locked on the land. Sea levels fell and land bridges linked the continents. The Central American land bridge, for example, opened the way to north–south land migration but effectively closed the connection between the Pacific and the Atlantic marine communities.

↑ **In the Miocene,** horses adapted to grazing on the expanding grassland environments. To avoid predators, they became fast, agile and alert. They grew larger, longer-legged and more like those of today.

↗ **The Wasatch Mountains** in Utah, USA, were uplifted by regional faulting in the Tertiary and then sculpted by glaciation. This snowy scene is a reminder that Earth is still in this glacial period, albeit a milder phase.

→ **The African sengi,** *Elephantulus intufi*, is a tiny placental mammal related to the elephant. It can bend and twist its trunk, using it as a tool to catch insects. The oldest sengis appeared in the Eocene.

Quaternary: the age of humans

The Quaternary period—divided into the Pleistocene and Holocene epochs—started with the onset of an ice age 1.8 million years ago that continues to the present. During the Pleistocene, ice repeatedly pushed out of the Arctic Circle into Europe and North America. The temperature drop associated with this advancing ice had a profound effect on life. The mammoth, rhinoceros, bison, reindeer and musk ox all evolved woolly coats to protect them from the frigid conditions. Intelligent hominids continued to evolve. As hominids proliferated on each continent, many of the giant mammals, flightless birds and reptiles disappeared. The Holocene began after the last major glacial advance 10,000 years ago. By this time, humans had spread throughout the planet, but their impact has been most marked since the development of agriculture. Present population growth, habitat loss and pollution may lead to an ecological collapse. It is now a race to see if global intellect and altruistic consciousness can rise quickly enough above selfishness to avert disaster.

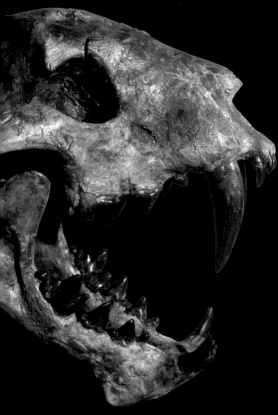

← **The saber-tooth** cat was one of the big carnivores of the Pleistocene that preyed on the large grazing herbivores. It became extinct shortly after human contact.

→ **During ice ages,** such as the current one, ice accumulates on polar continents and moves outward from them. The extent to which glaciers advance or retreat during an ice age greatly shapes the evolution of life.

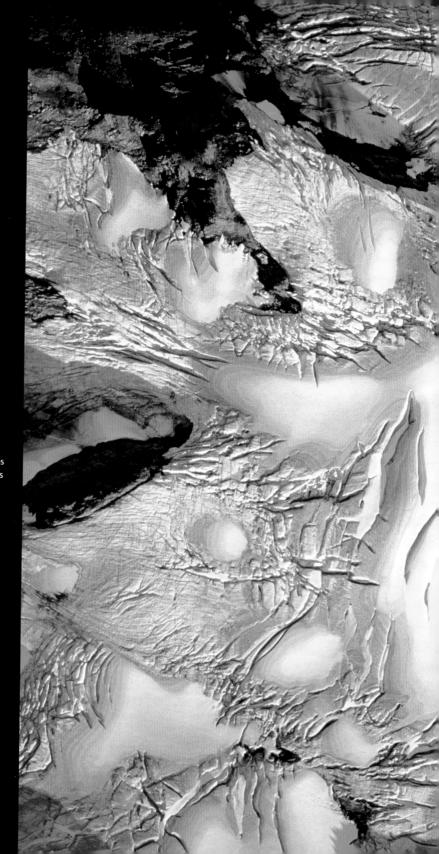

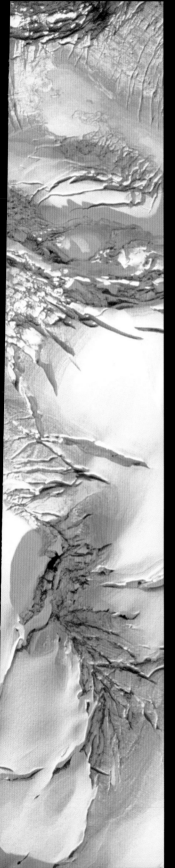

TODAY'S RESTLESS, COLD EARTH

The northward movement of India into Asia gradually forces the Himalayas skyward. Australia's northward journey causes subduction and volcanic activity in the islands of Southeast Asia. Africa continues to close the Mediterranean Sea. The Pacific plate is subducted all around its edges, causing a ring of volcanoes and earthquakes, while the Atlantic Ocean widens. Since the beginning of the Quaternary, ice on Greenland and Antarctica has stopped the oceans from flooding low-lying areas of the continents. During this ice age, barely noticeable eccentricities in Earth's orbit, called Milankovitch cycles, determine whether slightly more or slightly less snow accumulates on the polar landmasses in winter than can melt in summer. Over time, this seemingly insignificant effect causes glaciers to advance approximately every 100,000 years and then retreat. From Greenland, they can extend over the Arctic Sea and deep into Europe, Canada and the USA. Today, the world's glaciers are in retreat.

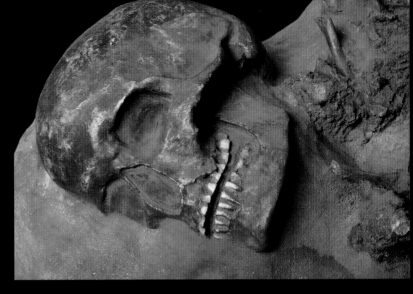

↑ *Homo sapiens* **skeletons** found at Dolni Vestonice in the Czech Republic reveal that anatomically modern humans lived in Europe 25,000 years ago. They made tools and ornaments, and painted cave art.

↓ **Mammal diversity** reached its peak during the Pleistocene. Since then, it has declined because of habitat destruction and overhunting. Here, Burchell's zebras and wildebeest graze on the grassy plains of Tanzania.

After the next extinction

After the Permian mass extinction, the reptiles rose to prominence; following the Cretaceous extinction, mammals seized the opportunity to fill their niches. This leads to the intriguing question of what will happen next. Which lifeforms might thrive after the next mass extinction? The possibilities are infinite. Perhaps highly adaptable, mobile species now considered pests, such as rats and weeds, will take over. Maybe species presently held in check by habitat loss associated with human population growth will thrive. It is envisaged that complex species, such as humans, will suffer. What would occur if we were greatly reduced in number? The next extinction may result from some catastrophic volcanic or meteor event, or perhaps from human-induced causes. Here, an artist muses.

THE FUTURE EXTINCTION SCENE?

Species numbers are down, perhaps at levels similar to those of the Triassic. This low biodiversity is due to the extinction event, combined with intense inter-species competition as the continents push toward another supercontinent reassembly. Much of the land is dry and hot. In the foreground, hungry carnivores crowd a mole rat colony broken open by a spiny rat with specially adapted claws and snout. Though the soldiers valiantly defend, the moles are no match for the spiny rat's tongue and the spine-wielding finches. Meanwhile, a fast-moving, long-legged crocodilian snatches a large flightless hen, despite its mate's attempts to ward off the predator with defensive sickle spurs and wing claws. A pair of grazing herbivores, with tusklike incisors for stripping bark, watch the scene.

Spiny rat, carnivorous placental rodent **I2**
Mole rat, hairless herbivorous burrowing rodent **K1**
Mole rat, soldier species with enlarged incisors **K0**
Giant rat, herbivorous capybara-like rodent **H4, K5**
Bufo species, large carnivorous toad **F1**
Crocodilian, long-legged, land-dwelling crocodile **D2**
Gallus futuris, evolved flightless fowl **A3, B1**
Finch, tool-wielding bird **H0, G2, K3**

Previous page Fossil crinoids appear as if still living,
waving in the ocean bottom currents of an earlier age.

Key features

Systematic study and classification of Earth's rocks and fossils is the key to understanding the bigger picture. Rocks continually cycle through Earth's mantle and are squeezed into massive mountain ranges. When unearthed, fossils reveal much about the time and place in which they lived.

How fossils form

Fossils form when an organism is buried and the hard parts of its body, such as bones, teeth, nails, shell or woody tissue, are preserved, either as original material or as an imprint. Sometimes the original material is dissolved away, leaving a cavity in the rock which may later become filled with another material, such as a mineral. The cavity is known as a mold and the internal filling as a cast. Fossils are common in fine-grained sediments that occur in low-energy environments such as lakes, seafloors, swamps and the quiet parts of deltas. They are rare in coarse-grained, high-energy sediments where they tend to be broken apart, abraded and destroyed. Hence fossils are mostly likely to be found in rocks such as limestone, chert, mudstone, shale and siltstone. A hammer and chisel can be used to split them apart along their bedding planes.

FROM PLANT TO FOSSIL
These autumn leaves and pine cones (*right*) falling into a lake have a good chance of becoming fossilized. Settling into the cold, oxygen-poor mud at the bottom of the lake, they will not decompose before they are buried by mud or silt. The result will be a carbonized imprint preserving every detail except original color. Silica replacement by mineralizing groundwater can preserve cell structure to the finest detail as in this fossil pine cone (*below*). Large buildup of vegetation in these circumstances forms coal deposits.

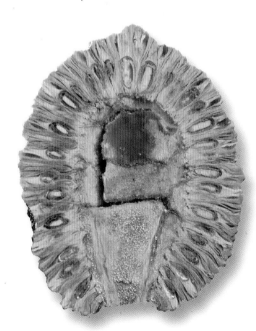

→ **Hard-bodied animals** have the best chance of being fossilized. Sea stars are protected by an external skeleton. Vast limestone deposits are made up of such fossils, including massive reefs built by colonies of tiny coral polyps.

HARD-BODIED VERSUS SOFT-BODIED ANIMALS

The vast majority of fossils are partial fragments of the preserved hard parts of an organism. Complete skeletons are extremely rare. Even rarer are the fossilized remains of soft-bodied organisms such as jellyfish. Countless entire families of such creatures have undoubtedly lived, evolved and died throughout the course of geological time without our knowledge. Precambrian organisms were entirely soft bodied, so the beginnings of life remain shrouded in mystery. Only rare localities, such as Mistaken Point in Canada and Ediacara Hills in Australia, provide a glimpse through foggy windows this far back. Here, entire Precambrian communities were buried extremely suddenly, such as by a catastrophic eruption of volcanic ash. If a soft-bodied organism is to be preserved and fossilized, its burial must be rapid and no oxygen must reach it.

STAGES IN FOSSIL FORMATION

Every living thing eventually dies but very few are preserved as fossils. This fresh dinosaur corpse lies on a lake floor, out of reach of scavenging predators.

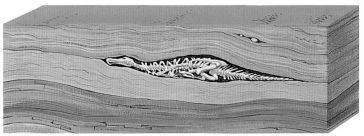

Burial must be fast to prevent the bones decaying or being washed away. The preservation of skin, flesh and internal organs is extremely rare.

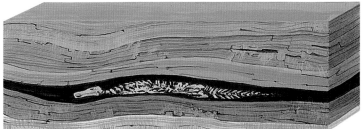

Compression and flattening of the skeleton occurs as layers of silt and sand build up on top of it. All the layers eventually turn to solid rock.

Uplift and erosion of the sedimentary rocks eventually release the bones from their stony tomb, often many millions of years after they were buried.

Types of fossils

Apart from the preservation of an organism's hard remains by way of burial in sediment, there are some less common ways in which fossils can form. Of these, petrification is the most frequent. Here, the original organic structure is impregnated by underground water containing dissolved minerals which solidify, literally turning it into rock. Usually the replacement is silica, sometimes in the form of precious opal. In such fossils it is uncommon to find any original material. Even fossilized shell material is rarely original, having been recrystallized and, in the process, losing its original sheen and luster. Preservation of soft body parts, skin, hair, scales and feathers is extremely rare, but there are some situations where this has occurred. Such preservation can occur when access of oxygen to the freshly dead organism is restricted. This halts the work of the decomposing bacteria. This may occur when an organism falls into the oxygen-poor waters of a bog or swamp, or is buried suddenly during a volcanic eruption. Among the more bizarre fossilizing mechanisms are entombing, freezing and drying. Animals, including humans and mammoths, have been preserved by being frozen in ice. Smaller animals, mainly insects, but sometimes lizards, frogs and birds, have become trapped and perfectly preserved in amber. Mummification is a rare fossilization process peculiar to desert areas.

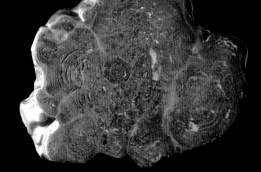

↑ **This tree trunk** has been fossilized by opaline silica that has replaced its cellular structure down to the finest detail, including its growth rings.

↓ **Unaltered shell material** of this Cretaceous ammonite shows its original iridescence. Such beautiful preservation is extremely rare.

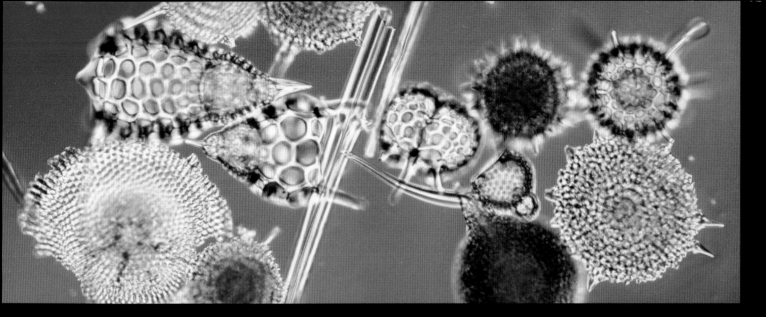

↑ **Radiolarian skeletons** are among nature's exquisitely beautiful miniature structures. These glasslike siliceous fossils of unicellular marine plankton were collected from Mt Hillaby, Barbados. Microfossils such as these allow geologists to link different parts of Earth in time and space.

← **Fossilized scales** of a bony *Dapedium* fish are preserved in minute detail in the mudstones at Lyme Regis, England. This fish died between 172 and 195 million years ago and was preserved in the seafloor mud of a deep tropical sea. Preservation of scales and skin in such fine detail is rare.

↓ **A spider's body** has been perfectly preserved when it became trapped in the sticky sap of a *Hymenaea* tree some 20 to 40 million years ago in what is now the Dominican Republic. When the resin hardens into amber, it becomes, like ice, one of the best preserving media.

Trace fossils

Some fossils are not actually remains of the organism itself. They are the marks, tracks, burrows, eggs, nests or other traces an animal left behind that became fossilized. While often difficult to assign a trace fossil to a particular species, these fossils may reveal more about lifestyle and habits than a fossil of the animal that made them. Footprints or feeding tracks left in soft silt or mud provide clues about movement. They may also show how an animal searched for food or how it captured other animals. Some fossil trackways reveal whether an animal traveled in a group or was solitary. Details about the manner in which it ran or walked may also be deduced. Diggings or burrows left by animals churning through seafloor sediment seeking food are distinct for different species and may reveal evolutionary development for groups as they developed more efficient feeding patterns over time. Fossil animal droppings, or coprolites, can be analyzed to find what type of food an animal ate and even how it chewed, swallowed and digested it. Rounded stomach stones (gastroliths) were used by some ancient animals to aid digestion, as do modern birds. The dental patterns of teethmarks on certain fossils may identify the predator. Nesting structures and eggs give insight into an animal's reproduction and how its young were reared. Sedimentary structures, such as ripple marks and glacial dropstones, reveal ancient water depth and temperature.

↗ **This nest of dinosaur** eggs was discovered in the Xixia Basin, Henan, China. Roughly hand-sized, these fossil *Hadrosaur* eggs are from the Cretaceous. This shows the unbroken underside of the hatched eggs in the nest.

→ **Coprolites are fossil dung;** these are from unidentified turtles and fishes and were preserved in Tertiary deposits. Coprolite contents often yield information about the animal's dietary habits.

↑ **Sedimentary structures,** such as these seafloor ripple marks in Glacier National Park, USA, when found together with trace fossils, yield information about the habitat in which an animal lived.

← **These parallel lines** are fossil tracks possibly made by now-extinct aquatic creatures, known as eurypterids, as they searched for food on an emerged sandbar in Western Australia, 420 million years ago.

Reconstructing fossils

Extracting fossils in the field, preparing them and eventually reconstructing the original animal is a delicate and time-consuming process. As in a crime scene, the position and orientation of fossil remains are important in ascertaining cause of death. The pieces of the puzzle must be carefully labeled and mapped before they are removed to make reassembly in the lab easier. Finding every possible piece is also important, as disarticulated fossils may be spread over a wide area—vertebrates can have over 200 bones. Fragile pieces may have to be stabilized with resin and protected with plaster before being moved. Back in the laboratory, reconstructing the original animal can take hundreds of times longer than collecting it. Depending on the nature of the fossil and its enclosing rock matrix, either mechanical or chemical preparation may be employed to extract the fossil. Chemical preparation is often used when the fossil is enclosed in limestone, which can be dissolved in weak acid, leaving the fossil behind.

→ **A great argus pheasant** feather shows complex fine structural detail. Fossils indicate that some dinosaurs were covered with primitive feathers. The evolutionary link between present-day birds and long-extinct dinosaurs was one of the most exciting scientific discoveries of recent times.

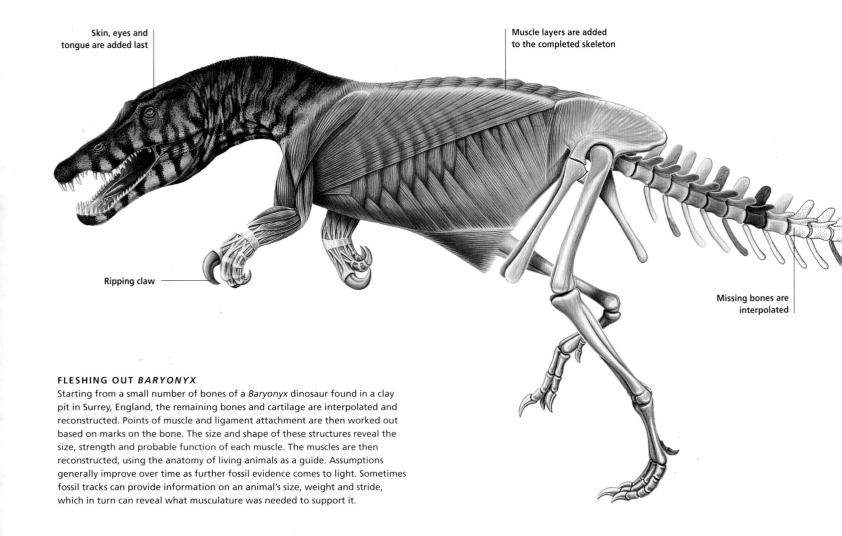

Skin, eyes and tongue are added last

Muscle layers are added to the completed skeleton

Ripping claw

Missing bones are interpolated

FLESHING OUT *BARYONYX*

Starting from a small number of bones of a *Baryonyx* dinosaur found in a clay pit in Surrey, England, the remaining bones and cartilage are interpolated and reconstructed. Points of muscle and ligament attachment are then worked out based on marks on the bone. The size and shape of these structures reveal the size, strength and probable function of each muscle. The muscles are then reconstructed, using the anatomy of living animals as a guide. Assumptions generally improve over time as further fossil evidence comes to light. Sometimes fossil tracks can provide information on an animal's size, weight and stride, which in turn can reveal what musculature was needed to support it.

THE MYSTERY OF DINOSAUR SKIN

Skin is the final and most visible stage of fossil reconstruction. Details are based on the study of fossils and close living relatives. In the case of dinosaurs, crocodile skin and markings are most commonly used. Recent discoveries of dinosaurs with primitive feathers at the Cretaceous Liaoning locality in China have created some alternative reconstruction possibilities. For example, *Deinonychus* (*above*), a small bipedal dinosaur, often depicted as lizardlike, may have been warm-blooded, birdlike and agile.

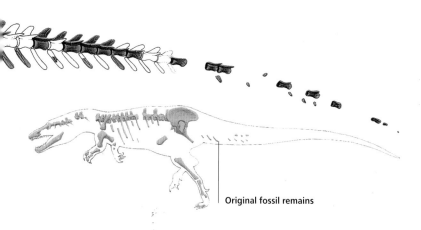

Original fossil remains

→ **The first step** in mechanical fossil preparation is to liberate the fossil from its enclosing matrix using an assortment of dentist-type drills and brushes—and a slow and steady hand. At the Black Hills Institute of Geological Research in South Dakota, USA, a paleontologist carefully prepares the skull of an *Acrocanthosaurus*, a large early Cretaceous carnivorous dinosaur.

Microfossils

Microfossils may be the smallest and least well known of the fossils but they are a vital tool for paleontologists. These tiny fossils help determine the time period and ancient environment in which a rock was laid down. Being so small and numerous they are often widespread and easily found. In the petroleum industry, microscopic hard remains of bacteria, protists (diatoms, coccoliths. forams), animals (sponge spicules, ostracods, conodonts), fungi (spores) and plants (pollen) are carefully extracted from drill cores by applying strong acid. These tiny fossils are identified and the information is used to correlate geologic strata from well to well, producing an accurate map of potential oil-bearing rock layers. Microfossils of various compositions can accumulate to form significant thicknesses of sedimentary rock. Limestone may be made up of foraminifera or coccoliths; chalk is from coccoliths; chert is from radiolaria and diatoms; and diatomite forms from diatoms.

INDEX FOSSILS

An index fossil is any fossil that can be used for correlating and dating geologic strata found in different parts of the world. A perfect index fossil will satisfy all the following criteria. It will have a short geologic range, so the time between its appearance and extinction is short. It will have a widespread geographic range, so it is found in many places around the globe. It will be found in various rock types, so it is not dependent on a particular type of bottom sediment. It must have fossilizable hard parts (either calcareous, siliceous, phosphatic or organic). It must also be extremely abundant so that it is likely to be found in even very small samples such as drill cores. Micro-organisms traveling the currents in the world's oceans (plankton) are excellent candidates for becoming index fossils.

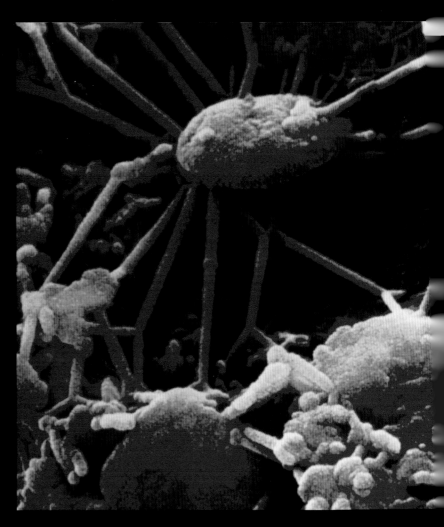

↑ **Bacteria fossils are among** the oldest known fossils, dated to over 3.5 billion years. The simplest lifeforms, they are prokaryotes—cells with no nucleus.

← **Foraminifera are single-celled** protists that either live on the seafloor or float in the water. Each constucts a tiny shell from calcium carbonate. Forams are particularly important index fossils.

→ **Fossil diatoms** retain their intricately patterned glassy silica shell cases. Called frustules, these may be either rounded or elongated. Living diatoms are single-celled, algae-like protists. They make up much of the plankton in the marine and freshwater food chains. Their shells accumulate by the millions on the seafloor, eventually fossilizing to form a siliceous sedimentary rock called diatomite.

Rocks and their minerals

Rocks are made up of one or more constituent minerals. They can be classified by studying the composition, proportion, shape, size and orientation of these minerals. Geologists study rocks in order to understand the processes—past and present—that formed Earth, and to learn where valuable and useful minerals and rocks are most likely to be found. Plate tectonics provides a regional framework for the process of geological exploration. The traditional method of studying rocks is to cut them into extremely thin slices and make microscope slides known as thin sections. The standard thickness is thin enough for most minerals to be highly transparent, revealing their internal structure and how they are intergrown with surrounding minerals. A special electron microscope, known as a microprobe, allows the operator to train an electron beam on selected minerals in a thin section and analyze the elements that make them up. Other techniques include using powdered rock samples to determine what percentages of valuable elements—such as gold, silver or copper—are present.

THE MINERAL WHEEL

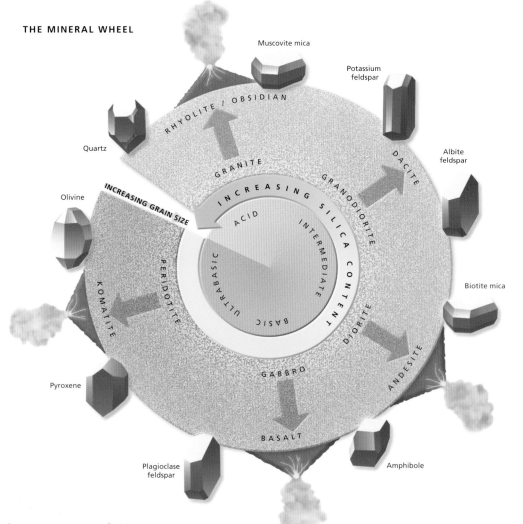

Marble is monomineralic—a rock made up of one essential mineral. A polished slab of red ornamental marble (*above*) and a specimen of its constituent mineral, pink calcite (*top*), are shown. Marble results from the heating and recrystallization of limestone.

THE MINERAL WHEEL

The mineral wheel (*left*) shows how nine common rock-forming minerals constitute the majority of the rocks crystallizing on Earth—the igneous rocks. Anticlockwise from olivine are rocks of increasing silica content. Rock-forming minerals are arrayed on the outside of the wheel to show the type of rock in which they usually dominate. In granite, quartz, muscovite, orthoclase feldspar and plagioclase feldspar are expected, but olivine will never be found. Peridotite, however, is made up almost entirely of olivine with some pyroxene. The wheel also represents Earth's crust: on the surface, rocks cool quickly, whereas beneath it the process takes much longer, and a coarser grain size can develop. Rocks of identical composition will have a completely different appearance and name as a result of their grain-size differences. These names are read off the wheel; for example, rhyolite is fine grained, while granite is the coarse-grained equivalent.

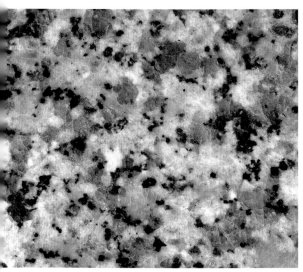

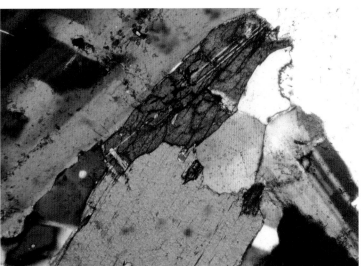

Rocks are studied at different scales. Granite creates a distinctive landscape in these outcrops on La Dique Island, Seychelles (*above*). A polished specimen (*far left*) reveals the minerals that make up granite—orthoclase (pink), quartz (gray), plagioclase (white) and biotite (black). The microscope-thin section (*left*) details how the mineral assemblage grew into a coarse, interlocking aggregate as the rock slowly cooled.

How rocks form

Rocks are constantly changing. Earth started as a molten ball, and the first rock to solidify into a primitive crust was basalt. Three rock groups form in distinct zones of the crust. Igneous rocks crystallize from molten magma, either within Earth or as it erupts as lava onto the surface. Metamorphic rocks are the product of other rocks that have been changed by heat and/or pressure. Sedimentary rocks are products of the breakdown of all other rocks. The rock cycle illustrates the way in which rocks may change from one form into another as a result of Earth's crustal processes.

THE ROCK CYCLE

By following the arrows around the rock cycle (*below*), it becomes evident that igneous rocks, born in spreading ridges, are recycled in subduction zones. There, heat and pressure change them into metamorphic rocks. These may eventually remelt, to erupt and recrystallize as andesitic igneous rock in a subduction mountain chain. Erosion of volcanic mountain ranges reduces rock into sediment, which gravity transports into basins. Here, it eventually compacts and hardens into sedimentary rocks. These rocks may, in turn, be crushed in collision zones where intense pressure causes them to recrystallize into metamorphic rock. Uplifted by this process, the forces of weathering and erosion again reduce the rocks to sediment. The rock cycle continues indefinitely and tends to concentrate the lighter elements at the surface. Consequently, as Earth evolves, there is a buildup of lighter continental crustal rocks.

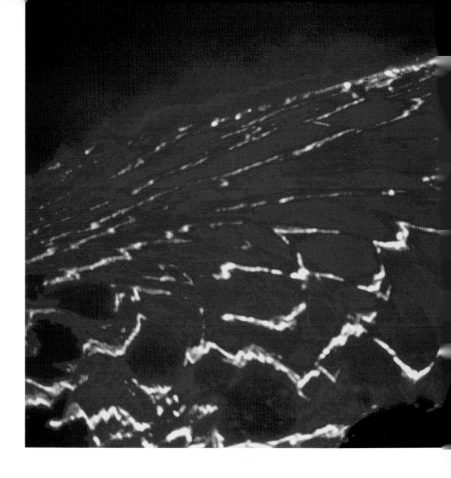

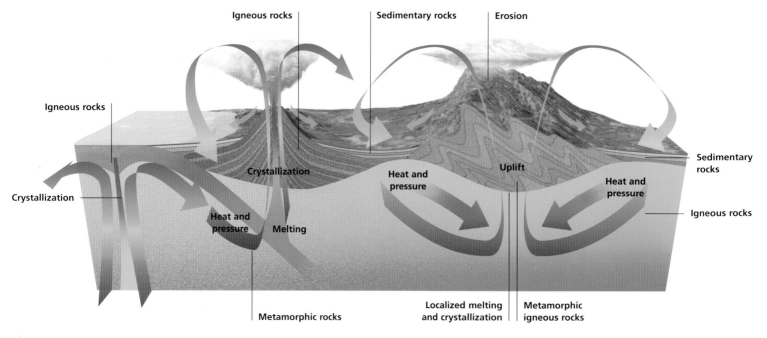

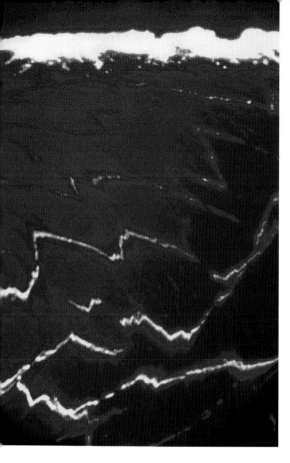

↑ **Igneous rocks** are born of fire, as seen in this lava lake in the crater of Hawaii's Kilauea volcano. As the lava cools and crystallizes, it develops a hard crust. This crust is constantly torn apart and shifted around by convection currents in the lava beneath. The processes here are highly reminiscent of those occurring on a massive scale in the crust, with the yellow cracks representing plate margins.

↗ **Metamorphic rocks** form under the strong horizontal compressional stress of continent-to-continent collision. In the Swiss Alps, this directional stress has pushed once-flat seafloor sedimentary rocks into a parallel series of mist-shrouded mountain ranges.

→ **Sedimentary rocks** will eventually be produced from the compaction and solidification of these loose sediments that make up the delta of the Colorado River, as it enters the Gulf of California, USA. The waters of the Colorado appear as a black dendritic web against the yellow–orange delta sands. The sands are deposited during times of flood.

Rocks from melts

Rocks that crystallize from molten magma or lava are called igneous rocks. Lava erupting for the first time along mid-oceanic ridges forms basalt. However, the more times a rock is melted, the more silica-rich or silicic it becomes. This process is known as fractionation and it occurs because silica-rich minerals melt at a lower temperature, leaving the silica-poor minerals behind. These silicic melts are lighter and accumulate at higher levels, forming thick continental crust. This crust is difficult to recycle into the mantle, so it is constantly deformed and remelted in continent-to-continent collisions.

CLASSIFICATION OF IGNEOUS ROCKS

Igneous rocks are generally classified according to two observable features—the size of their constituent mineral crystals and their composition. Chemical analysis of a rock determines how much silica it contains. Rocks with low silica, such as basalt and gabbro, are called basic rocks, while those with high silica, such as granite and rhyolite, are silicic rocks. Generally, silicic rocks are light in color as they are predominantly quartz and feldspar, while basic rocks are darker as they contain more iron and magnesium-rich minerals, such as pyroxene, biotite and olivine. The grain size of a rock depends on the rate at which it cooled.

Viewed through a polarizing microscope, these backlit thin sections of igneous rocks of similar composition reveal the effect of the cooling rate. The basalt (*below*) contains large, well-formed crystals of olivine that were already present when the lava erupted, but the majority of the rock is fine as it cooled quickly. The gabbro (*right*) cooled slowly, deep within Earth. It is comprised of coarse intergrown grains of feldspar and fractured olivines.

The granite terrain of the Devils Marbles, in Australia's Northern Territory, is typical of the balanced-boulder landscape caused by weathering along well-developed rectangular joints in granitic rock. Exfoliation, or onion-skin weathering, occurs when successive layers of rock are peeled from the surface of boulders, making them rounder and rounder. This peeling is caused by repeated heating and cooling, combined with chemical weathering along the resulting stress fractures.

IGNEOUS ROCK CLASSIFICATION

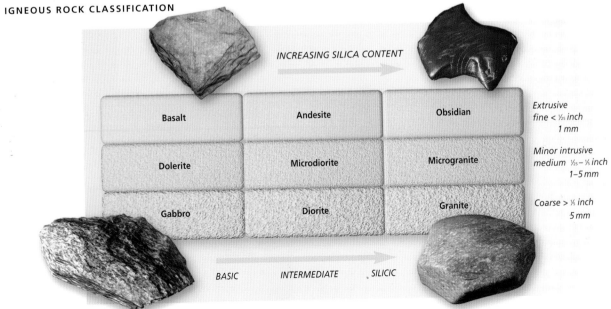

INCREASING SILICA CONTENT →

Basalt	Andesite	Obsidian	Extrusive fine < ⅒₅ inch 1 mm
Dolerite	Microdiorite	Microgranite	Minor intrusive medium ⅒₅–⅕ inch 1–5 mm
Gabbro	Diorite	Granite	Coarse > ⅕ inch 5 mm

BASIC → INTERMEDIATE → SILICIC

BASALT

This black, fine-grained basalt from Borambil, Australia, contains gas-bubble holes filled with a secondary zeolite mineral. The zeolite crystallized inward from the cavity walls some time after cooling. Basalt's crystals are quite small as it is a basic rock that cooled relatively quickly from a lava. Basalt is associated with mid-oceanic ridges and hot-spot volcanism. It has long been used as a building material and its fine-grained, even texture means it will take a uniform, high polish.

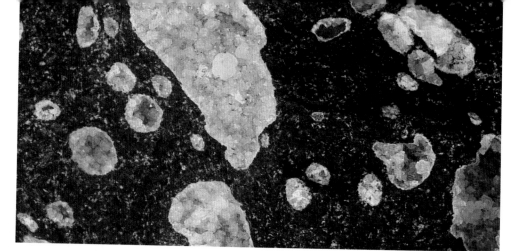

OBSIDIAN

This specimen of obsidian came from the aptly named Glass Mountain in California, USA. Obsidian is unique in that, although it is a silicate glass, it is non-crystalline. It has about the same composition as granite and rhyolite but, as a result of erupting and cooling quickly, its molecules did not have time to form an organized framework, so individual crystals did not grow. Slight movement in the cooling lava often causes flow-banding to develop. Obsidian played an important role in early American cultures, being used for knives, arrowheads and ornamental carving. It was even used for mirrors, because of its ability to take a brilliant polish.

PEGMATITE

This granitic pegmatite from California, USA, contains elongated prismatic crystals of black schorl tourmaline and equant red orange spessartine garnet in a white groundmass of quartz and microcline feldspar. Pegmatite grows in conditions exactly opposite to those of obsidian. It is a rock associated with the margins of granite intrusions. Cooling very slowly, usually in veins, its molecules have time to organize themselves into large, well-shaped, pure crystals. Pegmatite contains minerals made up of atoms that do not fit readily into the crystal structure of the normal rock-forming minerals. Hence many unusual minerals and metals are often found in pegmatites.

GRANITE

Coarse intergrown crystals of pink orthoclase feldspar, white plagioclase feldspar and gray quartz are clearly visible in this polished pink granite. At its center is an inclusion of fine-grained rock 2 inches (5 cm) across, torn from the magma chamber walls by the molten granite. Granite's crystals are coarse, as they cooled in massive bodies, or plutons, that lie well below Earth's surface and are exposed only through uplift and erosion. Granite is often found in linear belts associated with ancient subduction zones.

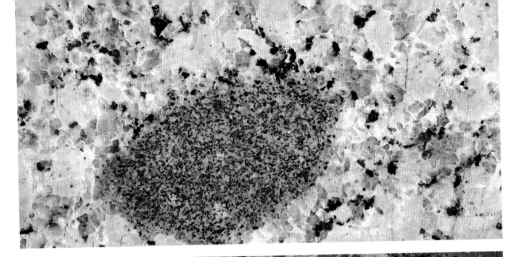

RHYOLITE

Rhyolite has the same composition as granite but, because it erupted onto the surface and cooled rapidly, the individual quartz and feldspar crystals of which it is comprised are not visible. Well-formed elongate crystals of feldspar in this specimen from Organ Pipe Cactus National Monument, Arizona, USA, indicate that some slow cooling took place before eruption. On eruption, these crystals oriented themselves parallel to the direction of flow and concentrated in bands. Rhyolite is associated with the melting of crustal rocks of granitic composition. If this lava contains a lot of gas, it froths and pumice forms, which can float on water.

SERPENTINITE

Serpentinite is made up of olivine and pyroxene that were altered and hydrated during intrusion. This specimen from England shows typical lamellar banding. This resulted from pressure alignment of its constituent platy minerals during squeezing and injection into Earth's crust. Linear belts of serpentinite that were squeezed up along fault planes, along with limestone and chert, represent zones of seafloor and mantle rocks that were crushed beyond recognition between colliding continents. Serpentinite belts are widespread in collision-formed mountain ranges such as the Alps and the Himalayas. This rock may be carved and polished as an ornamental material.

Rocks from heat and pressure

Metamorphic rocks have undergone a change in composition and texture. Heat and pressure associated with the plate tectonic convergent margin processes cause minerals to break down and reorganize their atoms into a structure more appropriate to the new higher temperatures and pressures. This takes place in the solid state. It starts when platy minerals, such as mica, begin to align themselves to the regional stress field. Then new, structurally robust minerals, such as garnet and staurolite, begin to crystallize. Metamorphic intensity increases until melting and crystallization take place and a new igneous rock is born as the rock cycle continues.

→ **Slate is often used** for roofing in the Grand Vonzo valley, Piedmont, Italy. Easily cleaved, physically strong and resistant to weathering, it can be inexpensively sourced here.

METAMORPHIC ROCK CLASSIFICATION

INCREASING TEMPERATURE

INCREASING PRESSURE

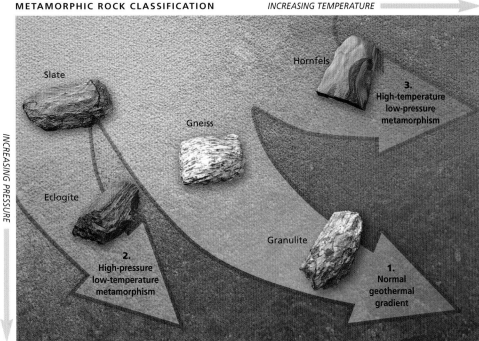

Slate

Hornfels

Gneiss

3.
High-temperature low-pressure metamorphism

Eclogite

2.
High-pressure low-temperature metamorphism

Granulite

1.
Normal geothermal gradient

CLASSIFICATION OF METAMORPHIC ROCKS

Starting at Earth's surface, this diagram shows the three different temperature and pressure pathways that the process of metamorphism can follow.
1. As depth increases, pressure and temperature rise. This is known as the normal geothermal gradient (*center arrow*). On this pathway, rocks change along its entire length, eventually turning into granulite before finally melting. Rocks formed by continent-to-continent collision also take this pathway because of high directional pressure in the zone of convergence at moderate temperatures. Slate, schist, phyllite, gneiss and amphibolite are found in these settings.
2. Subduction of cold ocean crust takes the lower pathway as the pressure in the crust increases much faster than the temperature can equilibrate (*lower arrow*). Increasing pressure produces slate, blue schist and eclogite (a metamorphosed basalt). Diamonds can also grow in these conditions. **3.** In volcanic belts, rocks in proximity to intrusive igneous bodies will be affected by high temperatures at relatively low pressures (*upper arrow*). Rocks forming along this pathway are known as contact metamorphic rocks and include hornfels (metamorphosed mudstone), quartzite (metamorphosed sandstone) and marble (metamorphosed limestone).

← **These quartzite outcrops** feature unusual zebra stripes. These are created by iron-bearing solutions migrating through the bedded rock. They are in the Uintas wilderness area, USA.

↓ **Two thin sections** of high-grade metamorphic rock show distinct foliation, or layering. Mylonite (*below*) exhibits layers of coarse recrystallized quartz and fine mica, while gneiss (*bottom*) has coarse, interlocking, white, gray and black quartz grains and elongate brown mica grains.

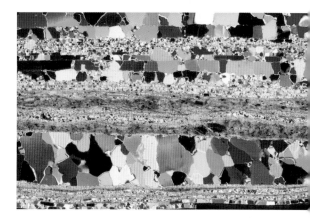

AMPHIBOLITE

Amphibolite is medium-grade rock formed by the regional metamorphism of basalt or gabbro. It is predominantly composed of the mineral hornblende, with plagioclase and minor quartz, epidote, ilmenite and magnetite. The elongated hornblende crystals are oriented by the pressure, giving the rock a shiny appearance when light is reflected off its surface. Amphibolite occurs in continent-to-continent collision zones and in Precambrian shields. This sample was found by the Bystraya River, Kamchatka, Russia.

GNEISS

This banded gneiss is cross-cut by a coarse white quartz and feldspar dike. A remobilized and recrystallized dike such as this indicates that temperatures and pressures were high enough to cause localized melting, injection and recrystallization. It was photographed at the Chola Pass, at an altitude of 17,700 feet (5400 m) in the Nepalese Himalayas— a continent-to-continent collision zone. Gneiss is a medium- to high-grade regional metamorphic rock consisting of alternating black micaceous schistose bands and coarse white granular quartz and feldspar granulose bands. It is formed by metamorphism of sedimentary rocks and may contain garnet.

PHYLLITE

Phyllite is a low-temperature regional metamorphic rock of an intermediate grade between slate and mica schist. It is composed essentially of quartz, sericite mica and chlorite, and is formed by the metamorphism of clay to sandy sedimentary rocks. The micaceous minerals align themselves at right-angles to the compression direction and give this rock its cleavage. This is less perfect than in slate because phyllite is more coarsely recrystallized. This specimen is from Nambucca Heads, Australia. Phyllite is found in continent-to-continent collision environments, often from retrograde metamorphism of rocks of a higher grade.

SCHIST

Schist is a medium- to high-grade rock formed by regional metamorphism of shales and arkoses. It is essentially quartz and mica—the quartz grows as a granular aggregate while the mica flakes align themselves at right-angles to the regional compression, producing a pronounced shine. Other minor minerals—such as garnet, andalusite, cordierite, kyanite, sillimanite and staurolite—grow as knots. Schist forms in continent-to-continent collision settings and in continental shield areas.

SKARN

Skarn is a high-grade contact metamorphic rock consisting of coarse granular wollastonite, calcite, pyroxene and garnet, often together with magnetite and sulfides (chalcopyrite, molybdenite, pyrite and sphalerite). Skarn forms as a result of hot, volatile-rich granitic magma coming into contact with impure limestones or dolomite. The volatiles allow the easy migration of elements, which recrystallize into coarse granular aggregates. Prospectors search for skarn mineralization around the margins of granitic intrusions and these are frequently mined for gold, copper, iron, manganese and molybdenum. This specimen is from the Boron Pit, in far eastern Russia.

SLATE

Slate results from the low-grade regional metamorphism of shale or mudstone. Metamorphism causes recrystallization and alignment of platy mica minerals at right-angles to the direction of the compression, which results in readily cleavable sheets. Being highly resistant to weathering, slate has been traditionally used as an inexpensive form of roofing and flooring. It has also been used for writing slates and blackboards, and for billiard tables, where weight and a high degree of flatness are essential. Small crystal knots are a feature of slightly higher grades. Slate is found at continent-to-continent collision boundaries, in the roots of old folded mountains such as the Appalachians in the USA.

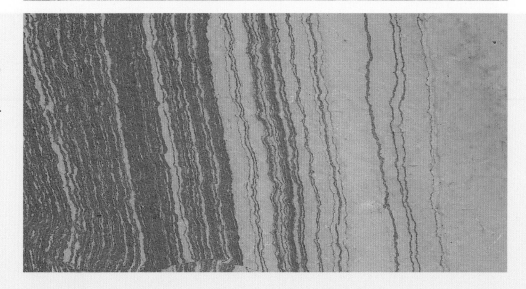

Layered rocks

Sedimentary rock is so called because it is formed from sediment. Sediment is material derived from the breakdown or weathering of pre-existing rock of any type. Erosion, water, wind and ice work with gravity to transport the sediment and deposit it in accumulations found generally at, or below, sea level. Huge deposits can occur along the flanks of growing mountain ranges forming at convergent margins or in deltas at the mouths of the rivers draining these areas. Coarse pebbles and gravels are deposited first. Then, as the energy drops, sand, silt and, lastly, the finest clays are deposited. Further deposition on top compacts these sediments, causing them to turn into rock, or lithify. These rocks are conglomerates, sandstones, siltstones and shales. In areas where there are no sediments, such as the deep ocean basins, organic or chemical sedimentary rocks can form. These include jasper, chert, ironstone, chalk and limestone. In swamps, coal results from the accumulation of tree and plant remains.

CLASSIFICATION OF SEDIMENTARY ROCKS

Sedimentary rocks are broadly classified according to whether they are formed from cemented mineral or rock fragments (clastic), from materials of organic origin, such as coral, shells or wood (organic) or precipitated from solution (chemical). The clastic rocks are named according to the size, shape and composition of the fragments they contain.

→ **Sandstone cliffs** at Navajo, in Arizona, USA, have beautiful red and yellow banding. This is not caused by sedimentary bedding, but by iron-rich water staining the originally white sandstone as it traveled through the porous rock.

↓ **The world's oldest** sedimentary rocks were deposited over 3000 million years ago, before the atmosphere had fully developed. Silicified yellow asbestos, or tiger eye, forms as bands in ironstone. It can be cut into an attractive gemstone.

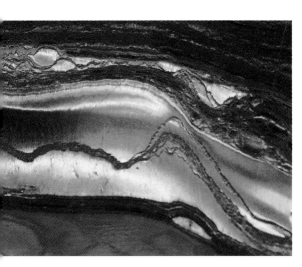

Breccia

Sandstone

Shale

Chert

	SEDIMENTARY ROCK CLASSIFICATION		
Particles	**Particle size: inches**	**mm**	**Sedimentary rocks**
Pebbles, cobbles, boulders	⅓ to 10	2 to > 256	Conglomerate, breccia
Fine to coarse sand	¼₁ to ⅓	1⁄16 to 2	Sandstone, arenite, graywacke
Fine to coarse silt	1⁄6506 to ¼₁	1⁄256 to 1⁄16	Siltstone
Clay	< 1⁄6506	< 1⁄256	Claystone, mudstone, shale
Chemical precipitates			Limestone, dolomite, evaporites, chert, bauxite

↑ **This backlit micrograph** of a thin slice of sandstone shows numerous semi-rounded quartz grains that have been cemented together into a quartz sandstone. Microscopic study of the composition and roundness of these grains enables geologists to determine where the minerals originally came from and how far they traveled. Rounder grains have traveled farthest.

← **Chemically deposited** rocks, such as these forming rims around thermal pools in Turkey, are part of the sedimentary rock family. The hot, mineral-rich waters coming to the surface precipitate many fine layers of calcium carbonate (travertine), along the raised edge. As the water flows over the edge of each terraced pool, these deposits create pretty, stalactite-like formations.

CONGLOMERATE

Conglomerates are lithified pebble and sand deposits formed in high-energy environments, such as floodplains, beaches and alluvial fans. In this specimen from Australia's New England region, round, waterworn pebbles of milky quartz are cemented in a gray, sandy matrix. These river deposits were mined for alluvial diamonds. Conglomerate pebbles may be fragments of any other rock or mineral and are rounded during transport. If the fragments are angular, the rock is called a breccia.

FLINT AND CHERT

Flint is workable nodules of chert found in the western European chalk deposits. A detail from a late Paleolithic flint hand ax from Farnham, Surrey, England, shows flint's typical conchoidal fracture. This property made flint useful to early humans as a material for making tools. Because flint produces sparks when two pieces are struck together, it was used as a fire starter and, later, to ignite gunpowder in firearms. Chert forms in marine deposits, either by direct chemical precipitation of silica or by the slow accumulation of siliceous microfossils on the seafloor, far away from sources of terrestrial sediments that would otherwise form sandstones and shales.

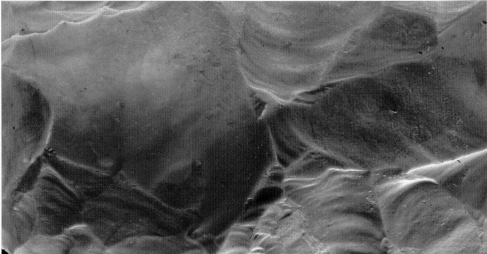

LIMESTONE

This polished limestone from Durham, England, shows white coral fossils in a fine-grained black calcareous matrix. Limestone occurs in shallow to deep marine environments, and, like chert, it can form as a chemical precipitate or from the accumulation of calcareous microfossils. It can also form spectacular, huge deposits from the remains of reef-building corals or calcareous algae. These may become formed into mountain ranges. Being quite soluble in water, the weathered surfaces of limestone often appear etched, and the landscapes of limestone terrains (known as karst) are characterized by underground rivers and caves.

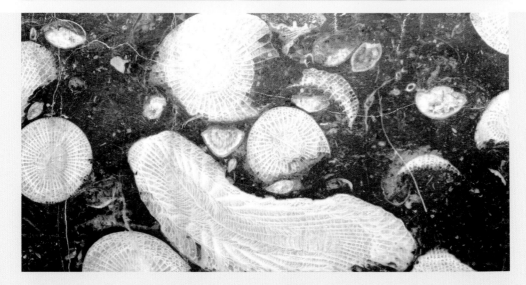

MUDSTONE

These dipping mudstone layers are exposed in the walls of Red Rock Canyon, Waverton Lakes National Park, Canada. The red tint is caused by oxidized iron. Mudstones and shales are composed of tiny, flat, clay particles that were deposited in horizontal layers in environments wherever constantly still water existed. These include lakes, low-energy floodplains and marine environments far from land and, therefore, sediment—the opposite energy extreme to that of conglomerates. Mudstones are used in brick and ceramic manufacture.

SANDSTONE

Sandstone, or arenite, is finer-grained than conglomerate and coarser than mudstone. Its particles are commonly rounded quartz grains but they may be feldspar and other minerals, such as mica. Graywacke is a sandstone composed of fine to coarse angular rock fragments that are usually associated with rapidly uplifting convergent margins. Sandstones form in moderately high-energy beach, river-delta or desert environments. Lithification binds the grains together with quartz, calcite or iron-oxide cement, and the nature and degree of cementation determines how hard the rock is. This specimen has been stained by groundwater containing iron oxide.

SHALE

This split sheet of shale from New South Wales, Australia, shows concentric zones of dark and light yellow bands due to iron-oxide-bearing groundwater seeping in toward the center from a rectangular network of cracks and fractures. Shale is finer than siltstone, as is mudstone. Both differ from shale in that they are composed of clay, which is made up of platelike particles and not rounded as in silt. Layering occurs because the flat clay flakes align themselves horizontally. Shale weathers easily so it is best seen exposed in fresh road cuts or canyon walls, where it is protected by layers of more resistant sandstones. Shale forms in low-energy environments.

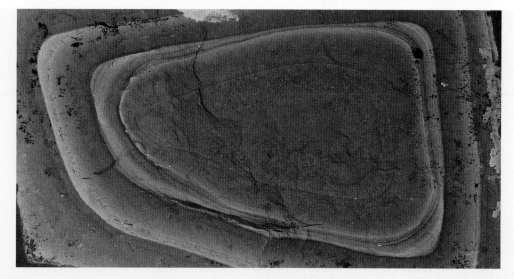

Previous page Viewed from space, a field of volcanoes
is eroded savagely by streams flowing down their flanks.

Earth's surface is a series of rocky landscapes that form in response to its internal heat engine. Lava and ash erupt from volcanoes, and mountains rise as continents crush together. Weathering and erosion work in the opposite direction: rain, rivers and wind reduce the level of the land to that of the sea.

...ones associated with fixed hot spots in the upper mantle. They are the gentle giants and are rarely explosive. People can watch red rivers of lava pouring out of vents day after day without risk to their personal safety. Composite volcanoes are usually quiet but, when they do explode, it is often violently because of trapped steam and other volatiles in the viscous lavas. These volcanoes arise from melting seafloor and sediments that are continually being pushed underneath the continental crust along the subduction zones. All the large, classic, steep-angle cones around the world, such as Mt Fuji, are composite volcanoes. Cinder cones, such as Paricutin in Mexico, are smaller and built up of rock fragments, which are explosively ejected from a vent.

↑ **The mid-Atlantic oceanic ridge** rises above the surface of the ocean and becomes visible where it crosses Iceland. In Thingvellir National Park, people can walk along the center of the ridge and view the cliffs formed by basalt lava flows. The flows have erupted from the fissure and, as the ocean continues to spread, they are constantly torn apart and filled by fresh lava.

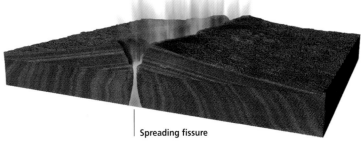

Fissure or rift volcano

Spreading fissure

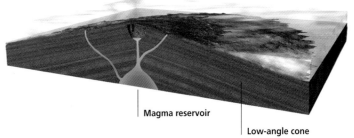

Shield volcano

Magma reservoir

Low-angle cone

Cinder cone

Steep cone

Feeder pipe

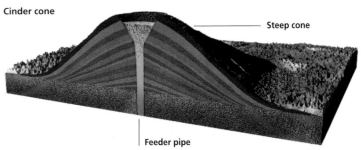

Composite or stratovolcano

Steep cone

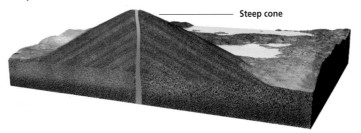

↑ **Rivers of fire** stream out of Puu Oo vent on Hawaii's Big Island and make their way to the sea. Shield volcanoes like this form very slowly, relentlessly, flow upon flow, year after year, until a broad, flat cone is built. It can rise to a great height above the seafloor. In Hawaii, this is more than 6 miles (10 km).

← **The perfectly formed** composite cone of Klyuchevskaya volcano in Kamchatka is Russia's most active volcano. Here, winter snow lightly covers its central summit and crater. Standing 15,900 feet (4850 m) above sea level, it is made up of alternating layers of pyroclastic debris (ash, cinders, blocks) and lava flows. Composite or stratovolcanoes grow in lines along active subduction zones. Eventually, these volcanoes can blow their tops catastrophically.

TYPES OF VOLCANOES
The four main types of volcanoes are shown here. Surprisingly, most of Earth's volcanic eruptions take place along the mid-oceanic ridges, but because they are almost entirely below the surface of the ocean, they go largely unnoticed. Composite volcanoes are steep-sided and are associated with subduction zones, while shield volcanoes are more flattened and occur at hot spots.

Eruptions

Volcanic eruptions are classified in increasing order of violence, with six styles, each named for a typical volcano displaying that eruptive style. The Hawaiian is a quiet outpouring of fluid basaltic lava. The Strombolian has lava which is slightly less fluid, with frequent small explosions releasing pent-up gases. The Vulcanian has less frequent but correspondingly more violent explosions as gas escapes from beneath a crust of solidifying lava. The Vesuvian, or Plinian, has violent explosions exemplified by the major 50-year cyclical eruptions of Italy's Mt Vesuvius. These clear out the old conduits and are then often followed by less violent eruptions. The rapidly built-up pressure of the gases trapped inside is released and, like soda fizzing out of a shaken bottle, turns the viscous rhyolitic lava into a mobile froth that pours down the mountain and solidifies as pumice. A Pelean eruption is the most destructive style. During this type of eruption, violent explosions generate *nuées ardentes* or glowing avalanches—hot, gas-charged clouds of magma droplets. These race down the mountainside and sweep away everything in their path.

↓ **Eruption of Mt Etna,** Italy, 29 July 2001. This satellite image shows the advancing reddish lava flows above the town of Nicolosi. Glowing brightly in the center is the summit crater and a smaller active fissure. Mt Etna is an active composite volcano built up of lava and ash.

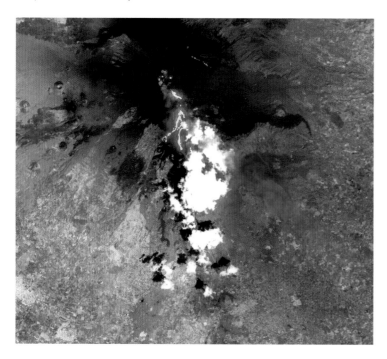

↑ **Lava eruptive phase** of Mt Etna, occurring as a continuous fire fountain of red-hot lava, builds up as a thick viscous mass around the top of the cone. The lava eventually flows slowly downward. These flows structurally strengthen the stratified cone of the composite volcano.

← **Explosive eruptive** phase of Mt Etna sees ash and dust shoot into the air. Most of this falls back to Earth and forms a blanket of debris on the sides of the cone, building it up. These blankets are soft and easily eroded.

Wind direction

Stratosphere

Troposphere

DUST AND ASH TRAVEL THE GLOBE

Plumes of hot gas, dust and ash from large volcanic eruptions can rise more than 6 miles (10 km) through the troposphere and reach the stratosphere. Here, strong winds can disperse the finer particles worldwide. This may even change weather patterns as the particle clouds reflect the Sun's rays back into space.

Ashfalls

Ashfalls can blanket huge areas, with the coarser fragments falling closer to the source of the explosion and the finer particles dispersing farther, sometimes worldwide. Blocks of over 100 tons (102 tonnes) have been thrown over 6 miles (10 km) by the the force of violent explosions. Ashfall is described as dust (fragments of rock, minerals or glass up to 0.2 inch or 0.5 cm); ash (from 0.2 up to several inches); and bombs, which are larger. These pyroclasts or fire fragments form volcanic deposits called tuffs and breccias. They are often used by geologists as valuable time-marker horizons because they are deposited everywhere simultaneously and decrease in thickness with increasing distance from the volcano. Large amounts of atmospheric dust particles can cause a dark volcanic twilight that will cool parts of Earth's surface by several degrees because they reflect sunlight back into space. Fallout and lower temperatures can kill plants and affect the animals that feed on them.

→ **Volcanic ash** blankets the houses and vegetation of a small fishing village on the Island of Saint Vincent in the Caribbean, creating an eerie, monotone gray landscape. Such dust and ash must be cleared off roofs and cars quickly. If it is left to build up, its weight will eventually crush the structure underneath. If it gets wet, this type of ash, which is made up of fine shards of silica glass and rock fragments, tends to set like concrete. No one was killed in this eruption.

↓ **Children living near** Japan's Sakurajima volcano wear hard hats on their way to school to protect their heads from flying debris ejected by the volcano. Dust and ash from the continual eruptions cover the streets.

LIFE IN AN ASHFALL ZONE

Edited journal excerpts reveal the true nature of life in an ashfall zone. "The first warning came from satellites tracking the ash cloud … and reports from cities where ash was beginning to fall. Soon we could see the cloud ourselves. Its leading edge was like a thick black curtain being drawn across the sky, blocking the sun … even though it was only early afternoon, it quickly became as dark as night. Hot gritty ash plastered our faces, clothes and hair, and our lungs hurt in the short dash to the house. We learned the trick of tying wet handkerchiefs over our mouths and wrapping the car's air filter with toilet paper." (Resident, Mt St Helens, 1982.) "At night, two 747 passenger jets unknowingly flew into the ash cloud. They plummeted nearly 25,000 feet (7625 m) due to loss of engine thrust but both pilots managed to restart their engines and land safely." (Air traffic report, Galanggung volcano, Indonesia, 1982). "This morning I swept the roof and gutters. Ash is building up on the mountainside now, and if the rain continues, deadly slides could occur. We may have to leave." (Resident, Mayon volcano, Philippines, 2000.)

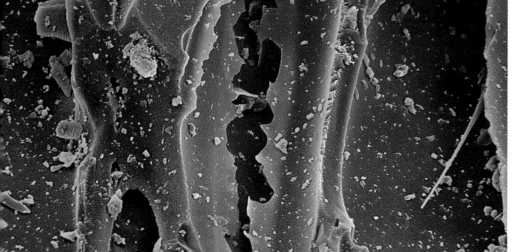

↑ **Red sunset** and sickle Moon provide a backdrop to the giant basaltic shield volcano of Mauna Kea, Hawaii. Fine volcanic dust particles pumped into the atmosphere by the 1991 eruption of Mt Pinatubo and dispersed by strong winds created this spectacular sunset worldwide.

← **A detail of a volcanic** ash fragment, which was created during the eruption of Mt St Helens in Washington state, USA. This scanning electron micrograph shows the extremely jagged edges that formed as a result of the explosive shattering. Hard, sharp shards of silicate glass such as this are extremely damaging to unprotected electrical and mechanical equipment, such as motors.

Craters and calderas

The idea of exploring the mysterious red hole at the top of an active volcano, the gateway into the planet's fiery interior, has long inspired scientists, writers, artists and poets. Volcanoes, fire and heat have become signature images associated with hell, demons and evil. This is not surprising, considering their destructive potential. Craters and calderas develop on the summits of most volcanoes above the eruptive vent. Craters remain when molten lava retreats into the magma chamber after the eruption has finished. In the case of explosive eruptions, it is the hole left after material is blasted out. Calderas, on the other hand, are much larger subsidence structures bounded by concentric ring faults, along which huge portions of the volcano's summit collapse into the empty magma chamber below. The cycle can begin anew with the birth of a small, central cinder cone, as seen at Crater Lake, in Oregon, USA, and continue until a majestic new volcanic cone again fills the entire caldera. Mt Fuji, centerpiece of Japan's Shinto religion, is a revered example—the five lakes that surround this volcano are all that remain of the former caldera lake.

CALDERA FORMATION STAGES

1 and **2.** A composite volcano forms above a shallow magma chamber that ejects ash and lava. **3.** Continued eruption depletes the magma chamber, leaving a void below the heavy volcanic cone. **4.** The magma chamber roof can no longer hold the weight of the cone and collapses along ring faults, leaving a flat-floored depression.

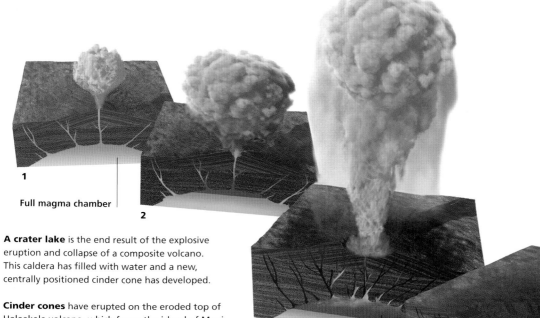

1

Full magma chamber

2

← **A crater lake** is the end result of the explosive eruption and collapse of a composite volcano. This caldera has filled with water and a new, centrally positioned cinder cone has developed.

↓ **Cinder cones** have erupted on the eroded top of Haleakala volcano, which forms the island of Maui, Hawaii. At the top of each cinder cone is a distinct, centrally located explosion crater.

3

Emptying magma chamber

4

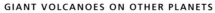

Main collapse fault

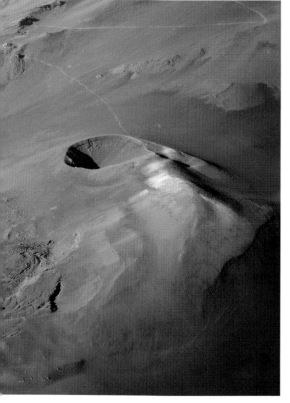

GIANT VOLCANOES ON OTHER PLANETS

Olympus Mons on Mars (*below right*) is the largest volcano in the Solar System. It is a basaltic shield 310 miles (500 km) across and reaches a height of 16.8 miles (27 km). At its summit, the clearly visible collapse caldera is 50 miles (80 km) across. Volcanoes occur whenever a solid crust encloses a higher-temperature molten interior. Jupiter's moon Io has volcanoes erupting sulfur, while those on Neptune's moon Triton spew forth icy methane. Below, Olympus Mons is compared to its diminutive relative, Mauna Loa, in Hawaii.

Olympus Mons
16.8 miles (27 km)

Mauna Loa
4 miles (6 km)

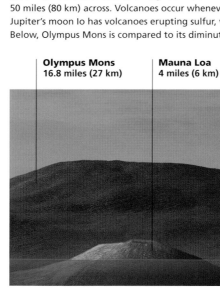

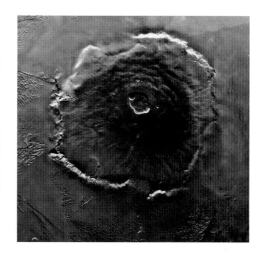

People have always lived near volcanoes, relying on their rich soils for agriculture, often oblivious to their danger. Today's population pressure means that, for many, there is often no other option. Interestingly, it is not the actual eruption that is the real danger—far more people die from secondary causes. When Tambora erupted in Indonesia in 1815, most of the 92,000 casualties suffered starvation due to crop losses. Others are swept away in explosion-generated tsunamis, as were most of the 36,000 killed when Krakatau exploded in 1883. Even today, preventable factors, such as disease and shoddy building practices, kill many—both were the case when 800 lives were lost in the 1991 Mt Pinatubo eruption in the Philippines.

1883 KRAKATAU ERUPTION, INDONESIA

Krakatau erupted in August 1883 as a series of four violent explosions followed by the collapse of the magma chamber. Two-thirds of the island was obliterated. The last explosion was heard a third of the way around the globe and threw so much ash and dust into the atmosphere that the surrounding Sunda Straits were plunged into darkness for a day. Giant waves reached heights of 130 feet (40 m) and destroyed 165 coastal villages. The waters of Sunda Straits were covered with a thick blanket of floating pumice blocks. Ash and dust were blown 30 miles (50 km) into the stratosphere and circled the globe. Apart from creating months of spectacular red sunsets, this acted as a solar radiation screen that lowered global temperatures by as much as 2.2°F (1.2°C) during the following year. Temperatures did not return to normal until 1888. Today, the very active Anak Krakatau (Child of Krakatau) is growing and refilling the caldera.

DEADLY VOLCANOES

Year	Volcano	Location	Death toll	Major cause of deaths
1815	Tambora	Indonesia	92,000	starvation
1883	Krakatau	Indonesia	36,400	tsunami
1902	Mt Pelee	Martinique	29,000	ashflows
1985	Ruiz	Colombia	25,000	mudflows
1792	Unzen	Japan	14,000	volcano collapse, tsunami
1783	Laki	Iceland	9000	starvation
1919	Kelut	Indonesia	5000	mudflows
1882	Galunggung	Indonesia	4000	mudflows
1631	Vesuvius	Italy	3500	mudflows, lava flows
AD 79	Vesuvius	Italy	3300	mudflows, lava flows
1772	Papandayan	Indonesia	3000	ashflows
1951	Lamington	Nuigini	3000	ashflows
1982	El Chichon	Mexico	2000	ashflows
1902	Soufriere	Saint Vincent	1600	ashflows
1741	Oshima	Japan	1500	tsunami
1783	Asama	Japan	1400	mudflows
1814	Mayon	Philippines	1200	mudflows

EARTH'S MOST DANGEROUS VOLCANOES

The most powerful volcanic eruptions are ranked here according to the number of lives lost. Clearly, the people of the Indonesian archipelago, an active subduction zone, have taken the brunt of volcanic activity.

← **The eruption** of Krakatau on 26–28 August 1883 received front page coverage in *The Illustrated London News* dated 8 September. The jagged remains of the once-symmetric cone, including the newly forming cinder cone, are depicted in the second illustration.

← **A few remnant** islands were all that remained after Krakatau exploded. The new cinder cone has black-sand beaches, and black lava flows have built out into the sea to the right.

1902 MT PELEE ERUPTION, MARTINIQUE

At 7.50 am on 8 May 1902, a huge plug of solidified lava was pushed out of Mt Pelee's throat, like a champagne cork being popped. A *nuée ardente* of gas and lava exploded through a crack in the plug. It roared down the volcano's flanks, destroying everything in its path, including all but two residents of Saint-Pierre. They had all stayed put, assured by the mayor that there was no cause for alarm—an election was due to be held three days later.

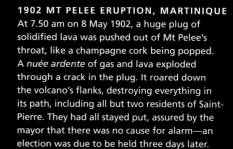

← **The village of Saint-Pierre** lies in ruins in the days following the eruption of Mt Pelee, which killed 29,000 people on the island of Martinique. The volcano is just visible in the background.

↙ **Although Mt Pelee** appears quite placid today, with its gentle green revegetated slopes, it is still an active volcano. Radial streams have carved into the soft, ashy flanks of the cone.

↓ **The people of Pompeii** were asphyxiated, then buried by ash from the AD 79 eruption of Mt Vesuvius, Italy. Archaeologists have reconstructed the scene by filling the spaces left by bodies with plaster, then excavating the ash to form molds.

Lava

When molten rock, or magma, spills forth onto Earth's surface it is called lava. It cools quickly, forming fine-grained or glassy volcanic rocks. Lava's characteristics depend on its silica and gas content, and the temperature of eruption. Basaltic melts (low silica content) erupt at above 2000°F (1100°C), forming fast-moving, highly fluid, bright red flows. Rhyolitic lavas (high silica) erupt at 1500°–1800°F (800°–1000°C) as slower-moving, more viscous flows. Carbonatite, an unusual carbonate lava, erupts at 900°–1100°F (500°–600°C) and looks like oozing black mud. It can be seen only at Ol Doinyo Lengai volcano in Tanzania. Advancing lava flows can fossilize whole forests in their path. The lava surrounds and carbonizes trees, then cools, forming a hollow mold that preserves an imprint of the original trunk.

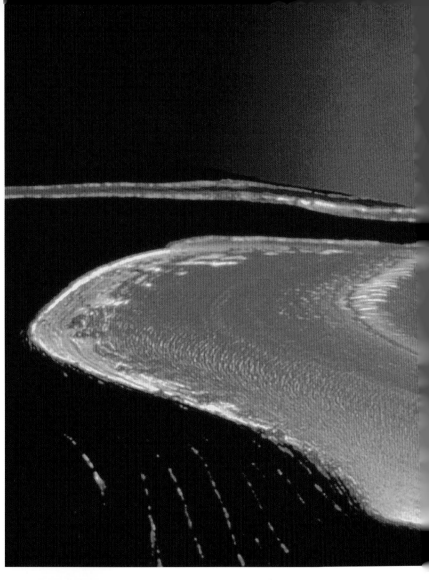

↑ **Thurston Lava Tube,** Hawaii, is the insulated conduit that allowed basaltic lava to stay hot as it traveled all the way down the flanks of the volcano to the sea. Like a garden hose, once the tap is turned off and the water drains out, all that is left is the empty, hollowed tube.

↑ **A red river** of highly fluid basaltic lava flows down a valley toward the sea. Eventually, the surface of this flow will cool and harden, while the river of molten material below continues on its journey to the coast. This is the first step in the formation of a lava tube.

← **A school bus** is partially buried by basaltic lava flows that inundated the village of Kalapana on Hawaii. Lives are rarely lost in these passive eruptions and the residents even had time to move their historic wooden church to a new locality higher up the hillside. The textured, ropy surface of the lava is known as pahoehoe.

↑ **Lava spills into the sea** continuously, flow by flow, building the coastline of Hawaii outward. The cold water shatters the lava into fragments that form the island's famous black sand beaches. This lava has the pahoehoe texture.

LAVA TUBE FORMATION
A lava tube forms where a basalt flow becomes hollow inside. The interior cools more slowly than the outside so a skin can form, leaving a hollow tube. This means the top of the lava becomes hard but the inside can still flow.

FORMATION OF A LAVA TUBE

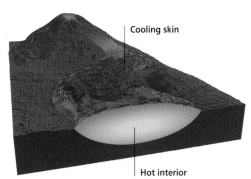

Cooling skin

Hot interior

1. Lava rivers quickly develop a hard skin shortly after the lava flow starts.

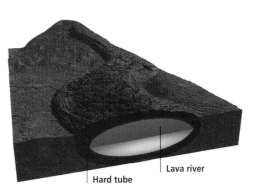

Lava river

Hard tube

2. Skin becomes harder and self-supporting, forming a tube through which lava still flows.

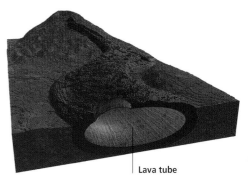

Lava tube

3. A hollow tube is all that remains when the vent stops supplying lava.

Columns, spires and dikes

Some of the world's most impressive landforms result from the erosion of volcanic terrains. Columns, spires and dikes are all resistant elements that have emerged in erosional landscapes due to more rapid weathering of the softer surrounding rock types. They are the ancient, hard, skeletal remains of now-unrecognizable old volcanic cones and their lava flows. The polygonal-patterned cracks in the tops of lava flows and pipelike cooling columns create cathedral-like scenery so unnatural that it is often ascribed to extraterrestrial activity or the gods. Columns usually have five or six sides and they range in size from 2 inches (5 cm) to 10 feet wide. When the columns are basalt, they are typically up to 50 feet (15 m) tall, while those formed from thick ignimbrite can reach up to 1000 feet (300 m). The shape can vary also, with horizontal surfaces usually producing straight columns, but those forming on slopes will follow the topography. Famous sites have names like Devils Postpile (California, USA) and Giants Causeway (Ireland). Organ Pipes is the most common name, featuring in Australia, Germany, Namibia and New Zealand. Columns are often used for building, as the pieces can be refitted like a jigsaw puzzle. Basalt columns were used to construct the 13th-century Somoska Castle in Slovakia, and the prehistoric Nan Madol people of Micronesia built structures using the columns like logs.

→ **Ship Rock** is a jagged volcanic plug with dikes radiating from it (one dike is visible in the foreground). Protruding from the floor of a desert in New Mexico, USA, this remnant is all that exists of a huge 30 million-year-old volcanic cone that has eroded away, as illustrated (*far right*).

↓ **A stairway of jagged columns** descends into the sea at Cape Raoul, Tasmania, Australia. Identical rocks in Antarctica show that Australia and Antarctica were once linked. Since the lava erupted and cooled some 55 million years ago, forming these columns, the two landmasses have since been torn some 1900 miles (3000 km) apart by Earth's tectonic forces.

COLUMN FORMATION
Lava cools from the outside in, toward the center, as it loses heat, both to the bottom surface and to the air with which the lava is in contact. Cooling leads to shrinkage of the surface of the lava mass.

Tension cracks, like tiny, three-rayed stars, begin to form at evenly spaced intervals over the entire cooling surface, which is under tension. The cracks grow toward each other and into the hot center.

Columns form when the cracks propagate through the entire mass and join up with one another. In a horizontal lava flow, the columns are vertical, as they grow inward from the top and bottom cooling surfaces.

SPIRES AND DIKES

1. In an active volcano, magma hardens in the internal plumbing of the volcano.

2. Erosion begins. Harder lava flows are undercut by the removal of underlying ash beds.

FORMATION OF SPIRES AND DIKES

1. When a volcano has finished its active life, any magma hardens in the internal plumbing of the volcano. This includes conduits such as the central vent (or volcanic neck), parasitic vents, radial dikes and sills. The processes of weathering and erosion soon begin to attack the now-dormant volcanic cone. Its layers of ash interspersed with lava flows are a soft target for radially draining streams and rivers that incise deeply into its flanks. **2.** Harder lava flows are undercut by the removal of underlying ash beds and they, too, soon collapse. Only a million years later the process is well underway, with the volcanic neck and radial dikes beginning to emerge. These, too, are worn away but, being hard and coherent, the process is much slower. Eventually, very little remains of the original cone, and the resistant neck and dikes stand up in bold relief above the landscape. **3.** The final diagram is a representation of Ship Rock National Monument, USA, as it can be seen today.

The volcanic neck of Devils Tower rises above the Wyoming plains, USA. The original molten plug of igneous rock has cooled into a series of columns, which, over time, were slowly exposed by the erosion of the original volcano.

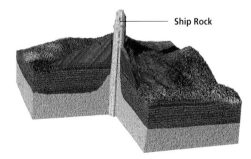

Ship Rock

3. Plugs and dikes are exposed, as seen in this diagram of Ship Rock National Monument, USA.

Geysers

Old Faithful shoots forth a column of boiling water some 150 feet (46 m) into the air every 76 minutes. Like a natural clock, it amazes the thousands of people who visit Yellowstone National Park in Wyoming, USA, each year. Geysers, essentially erupting hot springs, are rare phenomena because at least three vital natural conditions must be met for them to occur. There must be an abundant supply of groundwater, an intense source of heat close to the surface and a unique set of water- and pressure-tight plumbing. Over half the world's geysers are located in Yellowstone as the conditions here are exceptional—a hot spot exists below the area, and the rhyolite rocks provide abundant silica, which deposits a watertight seal on the walls of the geyser's plumbing. Geothermal fields are sometimes used as sources of energy. However, extraction can ruin geyser activity, as has occurred in Wairakei, New Zealand.

→ **Old Faithful** in Yellowstone ejects a column of boiling water into the air. A cone of deposited silica has built up around the opening.

↓ **Japanese macaques** relax in a volcanic-fed hot spring in Jigokudani Monkey Park, a favorite place to be during the long, cold winter. Magma melting from subducted crust produces the heat.

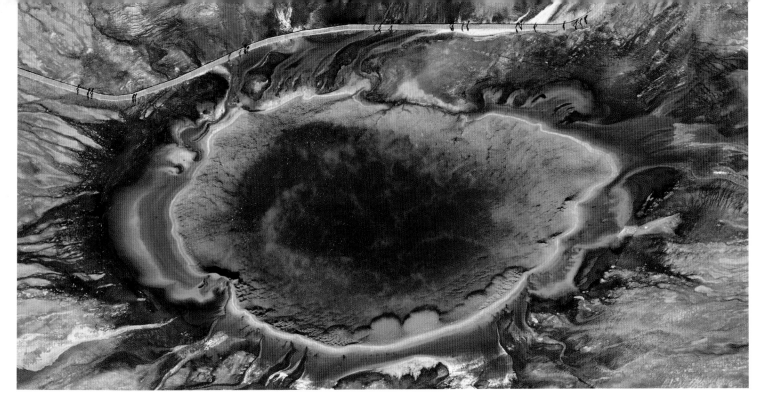

↑ **In Yellowstone's** Grand Prismatic Spring, microbial communities, each living at its own preferred temperature, create rings of color in the hot pool. The surrounding ground is warm, supporting grass that remains green even at sub-zero temperatures.

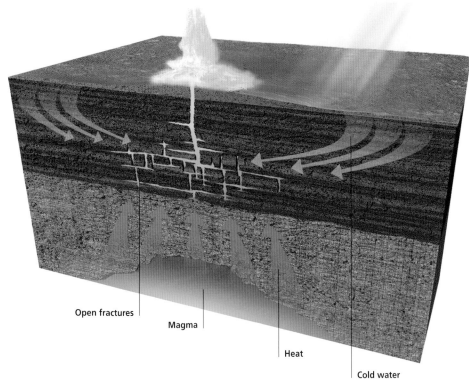

Open fractures

Magma

Heat

Cold water

THE MECHANICS OF A GEYSER

1. Cold underground water flows down into the plumbing system of a geyser, where it is met by heat and steam from below. The plumbing continues to fill while the water temperature rises to boiling point. **2.** Like a giant pressure cooker, the confining pressure of the overlying water allows the temperature to exceed boiling point. **3.** Beyond a certain temperature, the internal pressure exceeds the confining pressure and the system flashes suddenly to steam, which erupts from the vent as a boiling spout.

1   **2**   **3**

Deltas

Large rivers deposit enormous amounts of sediment at their mouths. Water spreads out from the confines of a river's channels and its sediment-carrying capacity drops. Coarser material being carried is dumped immediately, while progressively finer material is carried farther out to sea. The resultant triangular-shaped wedge of sediment is called a delta, after the Greek character. It is through the building, layer upon layer, of such sediment that sedimentary rocks are formed. The world's most powerful rivers, such as the Amazon, Congo, Yangtze and Orinoco, form delta deposits so immense that their weight causes depressions in Earth's crust. Early civilizations began on the fertile soils of deltas.

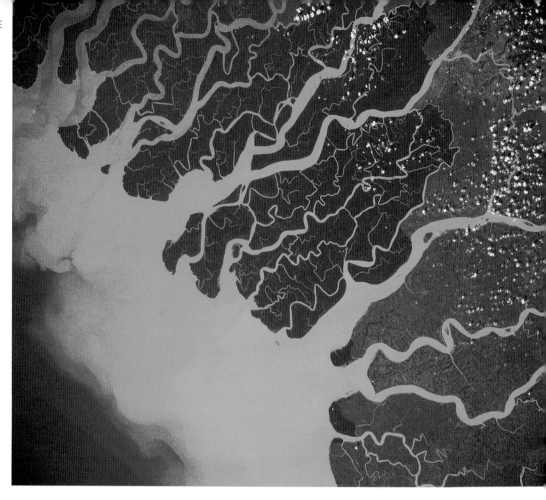

Satellite image of the Ganges River delta, Bangladesh, shows the sediment-laden water (light blue) of this mighty river entering the Bay of Bengal via hundreds of small meandering rivulets.

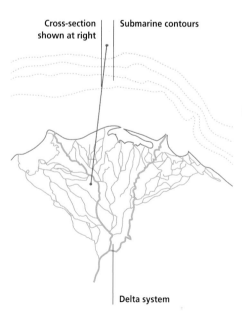

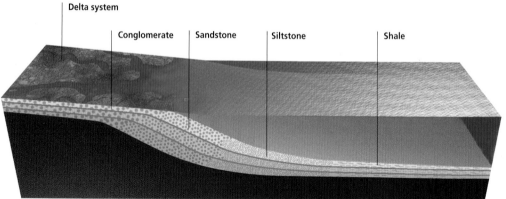

HOW A DELTA GROWS

As a river dumps each successive load of sediment, the delta builds out on top of earlier sediment. Only the part of the delta exposed above sea level is visible, but the true volume of sediment is enormous. As Earth's crust sinks downward under the immense weight, even more sediment is accommodated. Problems occur when projects cut sediment supply to the delta, such as in the Aswan High Dam on the River Nile. Sinking continues and the sea begins to flood and erode the delta.

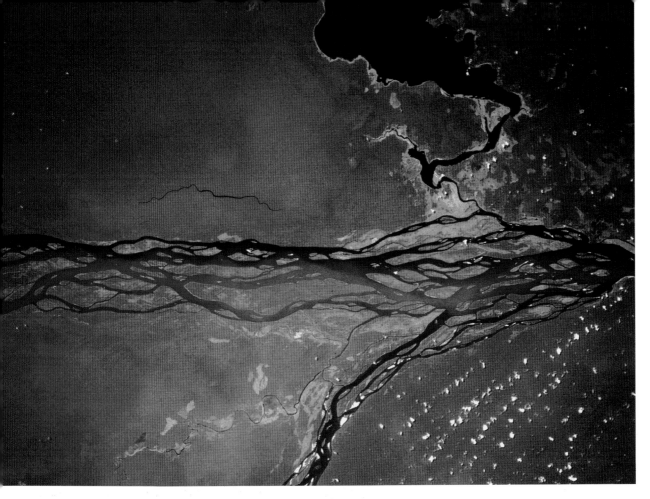

BRAIDED OR MEANDERING?

Two river systems are shown in the satellite images at left. Braided rivers are straighter, with multiple channels separated by sand or gravel bars. Following flooding periods, the river slows, choking with huge amounts of coarse sediment, and numerous small rivulets form rapidly. Meandering rivers are highly sinuous and confined to fewer channels that change position slowly. They carry finer sands and mud, and less sediment, than braided systems; therefore, they do not choke their channels.

The Congo, Africa's largest braided river, joins with the smaller Ubangi in a broad, flat valley on its journey to the Atlantic.

← **Glacial meltwater** carries silt into Peyto Lake, near Banff, Canada. Gray plumes show where the delta is actively building out.

WORLD RIVERS: POWER AND LENGTH				
River	Location	Discharge: yards³/sec.	m³/sec.	Length miles / km
Amazon	South America	235,000	180,000	4050 / 6516
Congo	Africa	55,000	42,000	2730 / 400
Yangtze	Asia	46,000	35,000	3960 / 6380
Orinoco	South America	37,000	28,000	1550 / 2500
Brahmaputra	Asia	26,000	20,000	1770 / 2840
Parana	South America	25,550	19,500	2800 / 4500
Mississippi	North America	22,990	17,545	3730 / 6000
Mekong	Southeast Asia	20,830	15,900	2800 / 4500
Ganges	India	19,650	15,000	1560 / 2510
Danube	Western Europe	8450	6450	1770 / 2850
Nile	North Africa	2080	1584	4160 / 6700
Euphrates	Middle East	1120	856	1750 / 2815
Ural	Russia	455	347	1570 / 2534
Colorado	North America	220	168	1450 / 2333

Waterfalls and riverbends

A waterfall is a sharp break in the long profile of a river as it slowly drops in elevation between its headwaters and its mouth. Any earth movement, such as uplift or faulting, will disturb the river's equilibrium and create a waterfall. It has long been recognized that waterfalls are constantly eroding, with the face moving back upstream. Given time, they will migrate for some distance, becoming shorter and shorter until they disappear completely. Pioneer American geologist James Hall established that Niagara Falls has receded 7.1 miles (11.4 km) during the last 12,300 years. So worrying was the rate of erosion that the US Army Corps of Engineers turned off the American Falls in 1969 by diverting its flow. They were then able to conduct stabilization work on the limestone rock face using rock bolts and steel cables. The rate of erosion at Niagara has also been slowed by the diversion of a large amount of water from the falls into pipes to generate hydroelectric power.

Riverbends are common where a river has lost most of its potential energy and meanders sluggishly within a broad alluvial floodplain or delta. They are typical of large rivers, such as the Amazon, that flow vast distances across continental plates with very little drop in elevation. Meanders migrate slowly by undercutting the outside of the bends and depositing on the inside.

→ **Victoria Falls** straddles the border between Zimbabwe and Zambia. Here, the Zambesi River thunders down into a narrow gorge across a rock face 1.1 miles (1.7 km) wide, dropping about 330 feet (100 m) to the bottom. The gorge follows a natural weakness in the basalt that gives the Zambesi the strange effect of emerging sideways from the base of the falls.

↓ **Incised goosenecks** of the San Juan River, Utah, USA, form some spectacular bends. Steady, slow uplift of the region has allowed this river to incise itself 1000 feet (305 m) into the sedimentary rocks of the region, while maintaining its original meandering pattern. Once incised, the meanders are no longer free to migrate about, as they were on the original floodplain.

Horseshoe Falls is one of three making up Niagara Falls. They are eroding upstream at a rate of about 3.3 feet (1 m) per year.

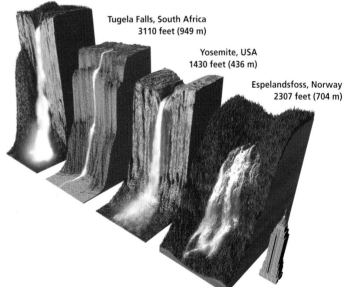

Angel Falls, Venezuela
3230 feet (985 m)

Tugela Falls, South Africa
3110 feet (949 m)

Yosemite, USA
1430 feet (436 m)

Espelandsfoss, Norway
2307 feet (704 m)

Empire State Building
1454 feet (443 m)

TALLEST VERSUS WIDEST WATERFALLS

Venezuela's Angel Falls, plunging 3230 feet (985 m) from Auyán Tepuy (Devils Mountain) in a single drop, has the honor of being the world's tallest waterfall. Other falls compete to be the widest, depending on exactly how they are measured (that is, including or excluding the dry parts in between). Iguazú Falls are 1.7 miles (2.7 km) wide; Victoria Falls are 1.1 miles (1.7 km); and Niagara Falls are 0.75 mile (1.2 km).

Gorges and canyons

Earth's deepest and most spectacular canyons are found in areas where the land is rising fast, such as collision and subduction margins, where rivers battle to cut their way down to sea level. The remote and rugged Yarlung Zangbo Canyon in Tibet takes first prize, with a maximum recorded depth of 3.3 miles (5.3 km). Nearby Kaligendege Canyon, Nepal, and Polungtsangpo Canyon, Tibet, are also giants, both with relief differences of over 2.5 miles (4 km). Phenomenal rates of uplift—over 1 inch (26 mm) per year in this region—are caused by the ongoing collision of India with the Asian continent. By comparison, the Grand Canyon, in Arizona, USA, seems small, with depths largely under 1.5 miles (2.4 km). Marine fossils, including trilobites, brachiopods, corals, mollusks and crinoids, are evidence of the canyon's marine origins. Some 20 million years ago, the uplift associated with the Rocky Mountains gave the Colorado River the energy to begin cutting the Grand Canyon. Many vast canyons are not visible—submarine canyons, cut into the continental shelves, were exposed only when sea levels were much lower.

THE GRAND CANYON

The Grand Canyon is the largest of the desert canyons, and probably the world's most visited. Its spectacular beauty is surpassed only by the immense expanse of geological time exposed in the canyon walls. The sequence spans nearly a third of Earth's history. The earliest rocks exposed in the base of the canyon are 1.7 billion-year-old schists, the roots of an ancient mountain range, while the youngest are only 250 million years old. The majority of the rocks seen in the canyon walls are limestones, sandstones and shales laid down in a shallow sea which, over geological time, sometimes rose and sometimes fell.

↗ **The Grand Canyon** is cut into a sequence of sedimentary rocks with the harder, more resistant rocks standing out as vertical-edged, cliff-forming units.

← **Yangtze Gorge,** China, is cut by the world's third-longest river. Ambitious damming projects for hydroelectric power and flood control have reduced this river's ability to erode and carry sediments to its delta and the sea.

→ **A bedded sandstone** canyon in Arizona, USA. The noon sun lights the floor of this dry desert canyon. Such canyons can briefly carry huge volumes of water during flash floods.

FORMATION OF A CANYON

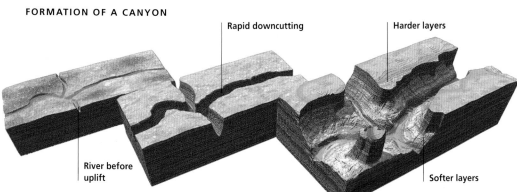

Rapid downcutting

Harder layers

River before uplift

Softer layers

1. CANYON FORMATION
Steep canyons form in resistant rock types such as well-cemented sandstone, but rivers must first be given their downcutting (kinetic) energy. This usually involves tectonic activity raising the land surface above sea level. The faster the uplift, the steeper and narrower the canyon will be.

2. CANYON DEEPENING
Here, rapid uplift has caused the river to incise a steep canyon in the topmost resistant sandstone layer. When the river cuts its way downward into an underlying layer of softer shale, the hard sandstone can be undercut, which enables larger blocks of this rock to break away.

3. CANYON WIDENING
Undercutting and collapse of the resistant sandstone continue as the river deepens and the valley widens at an accelerated rate. Resistant layers maintain a vertical valley wall while the softer layers develop a V-shaped profile. After deepening ceases, widening can continue.

↑ **Verdon Canyon** in Provence, France. This deep incision with very little valley widening belies the rapid uplift and resistant rock type of this canyon.

Mesas, buttes and tors

The world's surface plates are constantly moving; hence uplift and erosion occur all the time, slowly but continually. Mesas, buttes, pinnacles and tors are remnants that result from erosion on a massive scale. Vast quantities of rock material are removed to carve such landscapes—the original ground surface was at one time as high as the tops of these formations. The shapes of the erosional remnants depend almost entirely on the geology in which they form. The forces of erosion that shape these landscapes derive their energy from gravity; without plate tectonic uplift, the ultimate result would be an uninhabitable, marshy salt flat covering the entire surface of the globe within mere tens of millions of years.

MESAS, BUTTES AND PINNACLES

The term mesa is used when the width of the flat-topped summit is greater than the height of the escarpment. As the width of the summit reduces to the point where it is less than its height, the term butte is used. Finally, as undercutting of the caprock continues, only thin pillars of rock, called pinnacles, are left standing before they, too, finally collapse.

GRANITE TORS

Weathering of non-layered rocks such as granite (*left*) produces a landscape of rounded boulders or tors. Weathering begins along planes of weaknesses, when water seeping in during the day freezes at night. The expanding force of water turning to ice wedges off thin sheets of rock, progressively rounding and reducing the size of the boulders.

→ **Buttes and pinnacles** at Monument Valley, Utah, USA, are all that remain of a once-continuous plateau that linked their top surface.

FORMATION OF ROCK PINNACLES

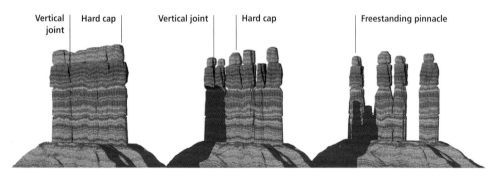

1. Thin cracks crisscrossing the rock are opened by water flowing down the hillside.

2. Freezing and thawing continues the widening process, creating an army of pinnacles.

3. The pinnacles fall, one by one, reducing the stone army, until none is left standing.

Deserts

The great deserts of the world include the Sahara, Kalahari, Great Indian, Atacama, Taklimakan, Gobi and the Australian deserts. These encircle the globe as belts between the latitudes of 20° to 30° north and south. This is where cool, dry air masses descend from the tropical convection cells and heat up, ensuring that rain rarely falls. Earth's counterclockwise rotation means that the surface winds here blow across the continents from east to west, creating the amazing desert-meets-ocean scenery along the western coasts of continents. Deserts can expand when large volumes of sand, normally stabilized by vegetation, are blown into massive migrating sand dunes that advance and bury anything in their path. Initially this was blamed on the overuse of marginal desert land for grazing and agriculture, but it is now becoming clear that humans are exercising a broader influence on global climate. This has come about through clearing vast areas of tropical forest, pouring greenhouse gases, such as carbon dioxide, into the atmosphere, releasing chemicals harmful to the ozone layer and damming or redirecting the flow of large rivers.

TYPES OF DUNES

Below are three different desert dune types: barchan (*bottom left*, Gobi Desert), seif or longitudinal (*center*, Australia), and star dune (*bottom right*, Namibia). Wind strength and direction are the factors that determine what type of dune will form. In a moderate-velocity uniform wind, barchan dunes will develop perpendicular to the wind with the horns pointing downwind. They migrate in the same direction. Uniform-direction, high-velocity winds will force the sand into the path of least resistance and produce seif or longitudinal dunes, which lie parallel to the wind. Star dunes form when the wind direction is variable—they are actually a series of barchan dunes at different orientations.

SAND DUNE MIGRATION

Dunes migrate when the wind becomes strong enough to roll individual sand grains. Sand is driven up the windward face to the crest, then cascades down the lee slope face. The grains are buried until they reappear on the windward face.

BARCHAN DUNE

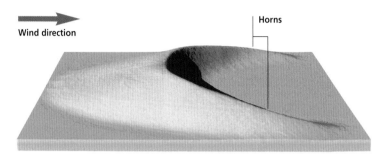

Wind direction

Horns

DUNE CROSS-SECTION

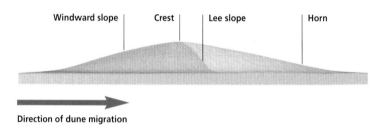

Windward slope | Crest | Lee slope | Horn

Direction of dune migration

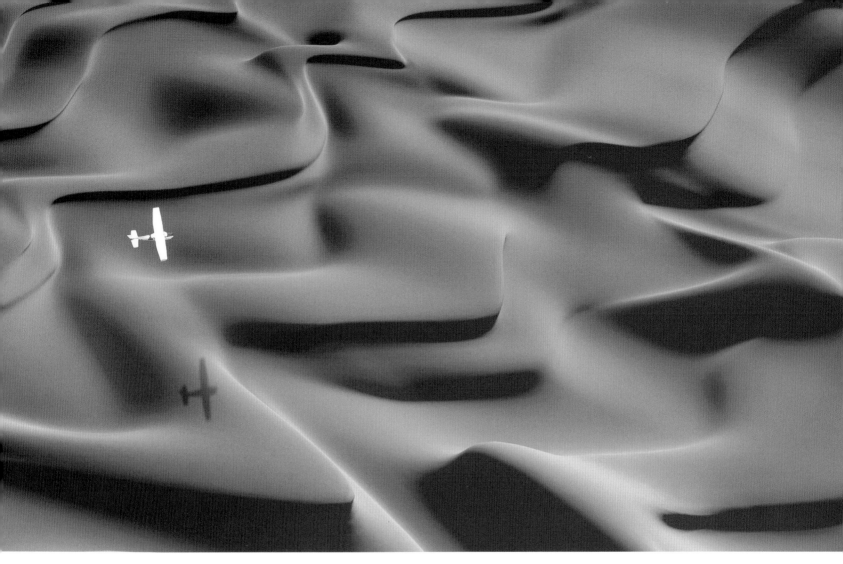

↑ **Giant dunes** in Namibia march across the desert floor, heading across the picture from top left to bottom right. The steeper lee slopes or downwind slip faces are clearly discernable (in shadow).

→ **Migrating dunes** can bury buildings when sand is not stabilized by vegetation. Poor landuse activity or climatic change may cause this.

↠ **Mummified** remains of people who inhabited the arid highlands of southern Peru some 2000 years ago have been found. Many of these fossils retain skin, hair and clothes that were naturally preserved by the extreme dryness of the Atacama desert.

Beaches and coastal remnants

The coastline—where the land meets the sea—evokes a wide variety of imagery, from the turbulent gray Atlantic pounding the foggy Scottish coastline to the warm, tranquil, turquoise lagoon of a coral atoll. Coastal scenery depends on the interrelationship of many factors. These include wave energy, tidal range, currents, ocean temperature, climate, vegetation, tectonic activity and, most importantly, coastline geology, such as jointing, layering or rock type.

HIGH-ENERGY VERSUS LOW-ENERGY COASTLINES

High-energy coastlines are subject to erosion by the ocean. This occurs mainly during storm activity, creating coastal cliffs and debris-strewn rock platforms. Low-energy coastlines are those protected from erosion by the open ocean with some form of offshore barrier. They are characterized by fine-sand beaches with low wave activity, estuarine swamps, quiet lagoons and river deltas.

→ **The shifting white sands** of a tidal channel in Australia's Whitsunday Islands means the scenery here changes by the hour as sand is continually moved and deposited by strong tidal currents that race through narrow channels between the islands. The six-hourly reversal in flow in such channels allows delta fans to develop at both ends of the channel, reminiscent of a giant bow tie.

↓ **Sea stacks, sea caves, blowholes** and arches may form under certain geological conditions. The Twelve Apostles, near Port Campbell, Australia, are sea stack remnants at the final stage before being entirely reclaimed by the sea.

CAVES, ARCHES AND STACKS

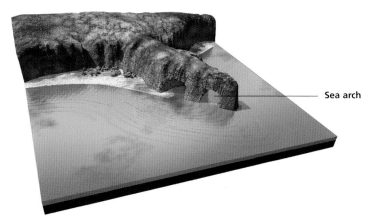

Sea cave

1. In the development of erosional coastal scenery, erosion along planes of weakness in the rock can open up a sea cave.

Sea arch

2. Caves become widened and enlarged, eventually becoming sea arches.

↓ **This glacial fjord in Norway** is typical of high-latitude coastal scenery. Here, enormous steep-walled, U-shaped valleys were reclaimed by the sea as glaciers from the last ice age retreated. The result is innumerable interconnected, navigable, scenic channelways.

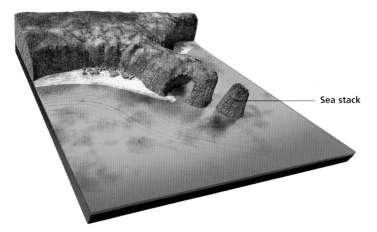

Sea stack

3. The top of the arch thins out and collapses, leaving a sea stack which, in turn, will eventually be consumed by the sea.

Karst

Rugged vertical karst peaks tower majestically over the tiny sailing junks in the Gulf of Tonkin in the South China Sea. Covered with twisted, stunted vegetation, such peaks have inspired a genre of Chinese and Vietnamese scenic artists. Karst refers to the type of topography that develops in areas of hard but water-soluble rocks, such as limestone. Features of karst are razor-sharp, boot-destroying, fluted and pitted bare-rock surfaces called lapies. Here, vegetation is starved of water as rivers rarely flow on the surface. Instead, they travel underground through a vast labyrinth of joint-controlled caverns. Mature karst terrains are typified by a series of residual peaks in advanced stages of erosion surrounded by flat land such as seen at Ha Long Bay, Vietnam. These are known as *hums* (Serbia and Montenegro), *pepino* or haystack hills (Puerto Rico), *mogotes* (Cuba) and stone forests or *fengling* (China and Vietnam).

HA LONG BAY, QUANG NINH PROVINCE, VIETNAM
Ha Long Bay (Bay of the Descending Dragon) was designated a UNESCO World Heritage Site in 1994. It is the world's largest and best-preserved mature limestone karst landscape. Karst is typified by tight clusters of conical, pointed mountains called *fengcong* (peak cluster) or flat plains with forests of singular peaks known as *fengling* (peak forest). The Ha Long karst is unique in having been invaded by the sea over a large area. Ha Long Bay contains some 1600 steep islands, mostly uninhabited. The stunning scene at right shows some of these islands. At water level, they display the characteristic marine erosion notches as the limestone is slowly eaten away. The Ha Long Bay karst formed over 20 million years.

KARST FORMATION

1. LIMESTONE FORMATION
Massive beds of limestone form by the continuous deposition of the tiny calcium carbonate skeletons of marine organisms living in a warm, shallow sea.

2. DEVELOPMENT OF SOLUTION CHANNELS
Elevated above sea level by tectonic activity, the limestone is subjected to rain and percolating groundwater, which begin to dissolve channelways.

3. STEEPENING OF THE KARST LANDSCAPE
Slightly acid rain continues to sharpen karst peaks and widen the channels into large valleys. Groundwater percolates into subterranean rivers and lakes.

← **Mature karst landscape,** such as seen at Ha Long Bay, on the coast of Vietnam, forms through a unique combination of slow uplift over millions of years under tropical climatic conditions.

← **Mature karst landscape,** such as seen at Ha Long Bay, on the coast of Vietnam, forms through a unique combination of slow uplift over millions of years under tropical climatic conditions.

↓ **The karst topography** of China features typical rugged terrain of very pointed peaks and wide, flat valleys. This *fengcong* landscape is represented in the third diagram in the sequence below left.

4. MARINE INVASION OF THE LANDSCAPE

Marine flooding due to the rising sea level, such as at Ha Long Bay, leaves a forest of peaks or *fengling* spectacularly protruding from the sea.

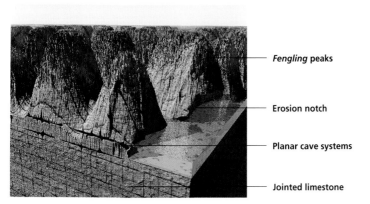

Fengling peaks

Erosion notch

Planar cave systems

Jointed limestone

Caves

Caves are solution channelways that have been dissolved by percolating groundwater. This water opens and widens the cave until entire river systems can flow through unseen from the surface. Underground rivers often enter the sea as turbulent jets of fresh water emerging through openings on the seafloor. Caves are intimately linked with areas of limestone karst, and their subterranean network may cover vast areas. Mammoth Cave system in Kentucky, USA, is the longest known, extending an amazing 346 miles (557 km), while Mexico's Huautla system in Oaxaca is one of the deepest, reaching 0.9 mile (1.47 km). When the groundwater level remains static for an extended period of time, extensive horizontal planar cave systems can develop. Circular doline depressions and sinkhole lakes develop when an overlying, thinning limestone roof finally collapses into the caverns below. These circular wells, known as cenotes, were considered to be "eyes into the afterlife" by the Mayan people of Mexico. Human sacrifices and offerings of jade and gold were regularly made to them.

Many caves are beautifully decorated in fairyland-like scenes created by calcium carbonate deposited on the walls, ceilings and floors by the same groundwater that dissolves the limestone elsewhere. Known as speleothems or cave formations, these include stalactites, stalagmites, helictites, shawls, pillars, columns and rimstone pools. Caving enthusiasts, called speleologists, love to explore and map cave systems, sometimes pushing the limit of their own safety with scuba expeditions through narrow submerged channelways.

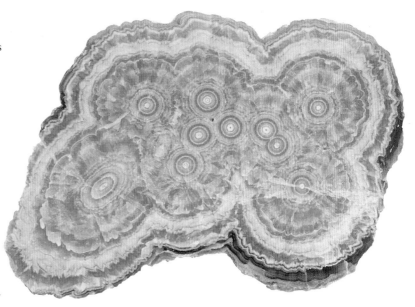

RHODOCHROSITE STALACTITE
Under certain conditions, cave formations can grow out of other minerals. In Catamarca Province, Argentina, a rare abundance of dissolved manganese in the groundwater has led to the growth of delicate pink-and-white banded rhodochrosite (manganese carbonate) cave formations.

LIMESTOME CAVE FORMATIONS

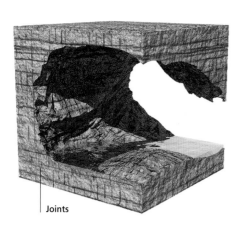

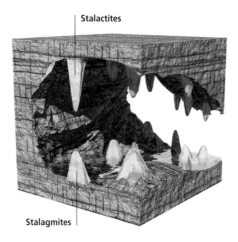
Stalactites

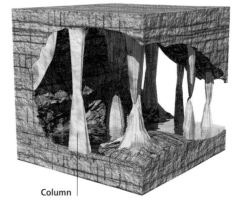
Column

Joints

Stalagmites

1. INITIATION
The same groundwater that dissolves the limestone to open a cave can precipitate it again as microscopic calcite crystals that sparkle on the cave walls.

2. STALACTITES AND STALAGMITES
Drop by drop, crystal by crystal, pointy stalactites form on the roof and more rounded stalagmites grow upward from the floor where each drip hits.

3. COLUMNS AND PILLARS
Eventually, pillars form when a stalactite meets a stalagmite. In time, these tenuously joined structures can fatten into substantial, thick columns.

The carbonate-rich, blue waters in Xpuhil cave, Dzitnup are part of the interconnected system of rivers and lakes that flow through limestone channelways and caverns beneath the surface of Mexico's Yucatan Peninsula. The pile of rubble, visible in the clear water, fell in when the roof collapsed, leaving a cenote or sinkhole through which daylight can enter. White calcite cave formations adorn the roof, along with thin tree roots reaching down from the dry surface above.

Swamps, bogs and tar pits

The words swamp, bog and tar pit usually conjure images of mosquito-ridden, smelly, sticky quagmires, but these are interesting, mood-evoking places that support a wide variety of wildlife. Such places are also excellent environments for preserving fossils. The fossil fuels on which we so heavily depend form in these environments. They all occur in low-lying areas where water is trapped, either permanently or seasonally. Swamps occur when areas are submerged by water on a regular basis, and the plants and animals they support are highly adapted to survive these conditions. Seasonal flooding of the Amazon and Orinoco rivers in South America inundates vast areas of rainforest. Coastal or estuarine tidal swamps are flooded with salt water on a daily basis, with the Everglades of Florida, USA, being the largest mangrove swamp in the world. Swamps act as giant silt filters for surface water running through them to the sea. Bogs, on the other hand, are more poorly drained and, being permanently wet, do not support larger vegetation. Tar pits are rarer as they occur only where hydrocarbons are able to escape from their reservoirs and rise to the surface along faults. Evaporation of all but the heaviest organic compounds (asphalts) leaves a series of sticky black pools that become practically invisible when covered by a thin layer of leaf litter. These pools are exceedingly dangerous for the wildlife living near them, as evidenced by vast numbers that become trapped and fossilize.

LIFE IN THE WETLANDS
Cypress Pond in Pine Log State Forest, Florida, USA (*right*), is home to a variety of specially adapted animals and plants. Adaptations to life in the wetlands include trees with extra roots above the oxygen-depleted zone; air roots for salt filtration; seed production timed for the non-flood season; floating seeds designed to reach dry land; and seeds that fertilize and germinate while still attached to the parent plant. Animals, such as beavers, have webbed feet for swimming and fine fur to trap air for insulation against the cold. Birds, such as herons, have long legs and beaks for wading and fishing.

The La Brea Tar Pits of El Rancho, Los Angeles, USA, are famous for the quantity and diversity of extinct Pleistocene animals. The pits formed around 40,000 years ago. Animals that became trapped in the sticky tar died there and eventually fossilized. Cats, bears, lions, wolves, sloths, mammoths, horses, bison, insects and plants have been excavated. Predatory saber-toothed cats were lured by an easy feast that was trapped in the sticky tar, only to become victims themselves. A laboratory and museum are on the site of the ongoing excavations.

A tannin-laden stream meanders its way slowly across the flat, treeless plains of a peat bog in Maine, USA. Nothing decays in this low-oxygen environment, so huge thicknesses of dead vegetation can build up below the living surface. After burial, this material becomes compacted and loses water. It then slowly goes through the coal-forming process. First, peat is formed, then it becomes brown coal, black coal and, finally, anthracite.

Tollund man is a well-preserved bog body dated at 240–220 BC. This 30 to 40-year-old man's mummified body was discovered in 1950 in a bog at Jutland Mose, Denmark. He died by being hanged using a braided leather noose. The excellent preservation is due to the low temperature, low oxygen levels and high tannin in the bog environment. The high tannin also causes such mummies to turn black.

Ancient seabeds and lakebeds

Many important sedimentary rock types form when sediments laid down on the floors of lakes and oceans turn to rock. Predominant are the huge volumes and thicknesses of silt, sand and gravel deposited as deltas at the mouths of large rivers. These will eventually become sandstones and shales. The coarsest sediments are dropped closest to land, with progressively finer sediments being dropped farther out to sea. The finest-grained rocks form far from the reach of land sediments and are made up of truly marine sediments laid down in deep ocean basins. In these areas, nothing appears to accumulate. Even after a lifetime, there is little more than a light dusting of sediments, as can be observed on deep oceanic wrecks such as that of the *Titanic*. However, over many millions of years, skeletons of microscopic marine animals and plants slowly but constantly rain down onto the ocean floor. Eventually, they build up significant deposits, such as the famous chalk cliffs exposed by erosion along the English and French coastlines. Micro-organism accumulations will form different rock types depending on their composition. Calcareous coccoliths (marine plants), ostracods (bivalved crustaceans) and foraminifera (protozoans) can build up chalk and limestone deposits, while deposits of siliceous diatoms (single-celled freshwater and marine plants) and radiolaria (protozoans) will form diatomite, flint, chert and jasper. However, these rocks do not always form from these tiny skeletal remains. They can also be chemically deposited, as occurs spectacularly in limestone caves and around hot springs.

↗ **The prow of the *Titanic*** lies on the Atlantic Ocean floor, several miles deep. The ship sank in April 1912, after hitting an iceberg. Apart from some marine growth, very little sediment has settled on the wreck.

→ ***Ichthyosaurus* was a dolphin-like creature** that lived in ancient seas but perished before the dinosaurs, some 66 million years ago. This fossil in slate includes an infant and five unborn young. Some ichthyosaur specimens are so perfectly preserved in seafloor sediments that they may yield stomach contents, skin and even body tissue. From such specimens, it can be deduced that these animals gave birth to live young at sea.

⇉ **The town of Marble Bar,** in Western Australia, is named for an extensive red jasper outcrop laid down in Archaean seas, 3.5 billion years ago. It was later tilted and folded. Such rock reveals the conditions in the ancient seas at that time because this type of rock forms continuously in deep oceans and contains traces of the elements that were laid down with it.

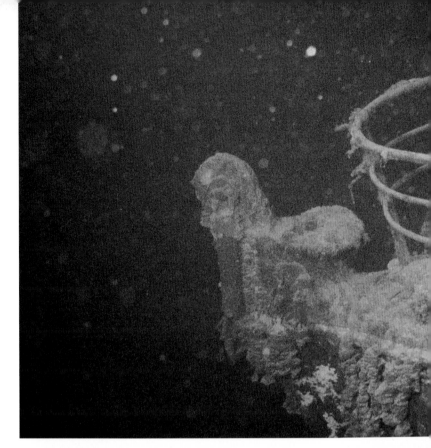

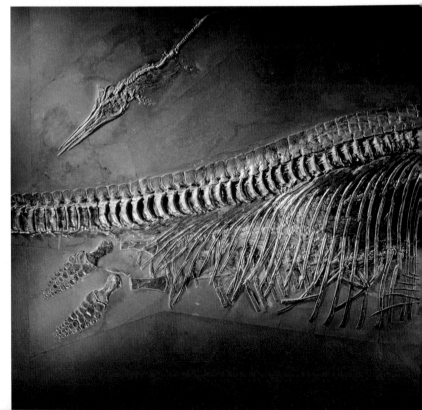

Fossil sites

Very rare and special conditions are needed to preserve high-quality fossils, particularly the soft-bodied forms, but over life's 3500 million-year span these rare conditions have occurred a number of times and in a number of places. Some rich fossil sites have been discovered. For instance, at Mistaken Point, Canada, and in the Ediacara Hills, Australia, bizarre Precambrian soft-bodied marine lifeforms, the oldest yet discovered, have stirred much speculation about early life. The Burgess Shale, Canada, spotlights the Cambrian explosion of life as it took place on a deep seabed. In Germany, a community of sea stars, trilobites and other unusual arthropods and cephalopods, known as the Hunsrückschiefer Fauna, provides a rare window into the Devonian period. Their soft parts have been perfectly replaced by pyrite. At Mazon Creek, USA, a mixed marine and terrestrial river delta community from the Carboniferous has been very well preserved. Permian–Triassic fossil beds at Karoo, South Africa, witness the transition from reptiles to early mammals. Holzmaden and Solnhofen in Germany have yielded rare complete skeletons of ichthyosaurs, plesiosaurs and pliosaurs. Other significant fossil sites are the Mesozoic dinosaur sites in the United States, and the Quaternary ancestral human sites in Africa.

↗ **River channel** and delta sandstones and shales of South Africa's Karoo Basin contain an excellent fossil record of the largest and near-total mass extinction event that occurred at the end of the Permian.

WORLD'S OLDEST FOSSIL SITE

At 565 million years old, Mistaken Point, Canada, is the oldest assemblage of complex multicellular organisms yet found. The fossils were preserved as impressions on the underlying rock surface when ash from a volcanic eruption buried the living community. Deep-seafloor fossils include frondlike leafy forms, cabbage-shaped growths, treelike networks, long spindle shapes and disks. Many are incomparable with any living animal or fossils found in other Precambrian sites.

← **World-class fossil** sites span the globe and geological time. Major sites yielding abundant fossils provide vital clues about Earth's evolutionary ancestry.

Dinosaur bones are being excavated from 150 million-year-old Jurassic river channel sandstones at Dinosaur National Monument, in Utah, USA. In this single quarry, the remains of nearly 100 individuals, representing 10 species, have been found. Most are types of long-necked dinosaur, such as *Diplodocus*, but there are other plant eaters as well as meat eaters, such as *Allosaurus*. This abundance is the result of bodies collecting in a sandbar as they were washed downstream in a huge river.

Folds and faults

Folding and faulting occur along all Earth's moving plate boundaries. At plate collision margins, rocks are crushed and broken. This is where compressional faulting—also known as reverse and thrust faulting—is common. As the time involved is so long, and the forces so constant in these zones, hard, brittle rocks behave more like soft plastic as they are pushed into folds like a crumpled rug. The tops of the folds (hills) are known as anticlines and the troughs in between are called synclines. Along zones where the crust is being pulled apart or being domed upward, tensional faulting (normal faulting) is the norm. This forms rift valley landscapes, examples of which are the Icelandic and East African rift valleys. At boundaries where plates traveling in opposite directions grind continuously against one another, strike-slip or transcurrent faulting is the most common, with the San Andreas fault, USA, being a famous example. The Great Glen fault, crossing Scotland from coast to coast as a straight line of lochs, is another spectacular example. In geological time, the sliding is slow and continuous but in real time it occurs as a series of short, sharp jumps, each one releasing earthquake energy.

→→ **The San Andreas fault,** from the air, is visible as a huge scar across the surface of Carrizo Plain, on the west coast of the USA. As the fault is in constant motion, everything crossing it soon becomes torn in two and displaced. This includes rivers, roads, fencelines, pastures and even buildings.

↓ **This synclinal fold** is in a mountainside in the Canadian Rockies. The sedimentary layers are highlighted by drifts of snow. Over vast periods of time, an enormous amount of rock has been removed from the once-surrounding, even taller anticlines.

FOLD STAGES

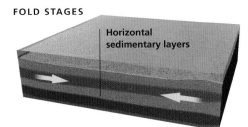

Horizontal sedimentary layers

1. The original flat-lying layers are uplifted and folded by the mountain-building forces.

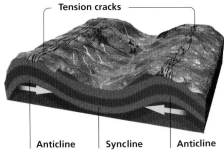

Tension cracks

Anticline   Syncline   Anticline

2. These become eroded. The tops of the anticlines go first as they have been weakened by the folding.

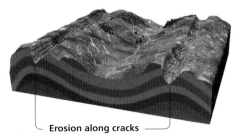

Erosion along cracks

3. Streams localize on the eroded anticlines and gouge out rapidly forming, deep valleys.

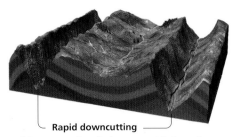

Rapid downcutting

4. This process is called topographic inversion, as valleys become mountains and vice versa.

TYPES OF FAULTS

Normal faulting occurs when rock fractures under tension and the layers are offset along a planar surface that has an angle of 45° or greater.

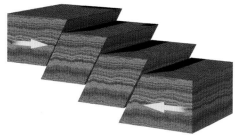

Reverse faulting occurs when the rock fractures under compression and is displaced along a planar surface in the reverse direction to a normal fault.

Thrust faulting is similar to reverse faulting but the fault plane angle is shallower (less than 30°), so layers are pushed farther on top of one another.

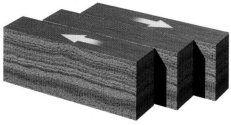

Transcurrent faulting is defined as horizontal or strike-slip offset along an essentially vertical fault plane, such as the San Andreas fault.

Glaciers and icefields

Icefields form when winter snowfall exceeds the amount that melts away in the summer. As the dome of ice builds, it pushes ice outward at its edges, forming glaciers that advance slowly into warmer areas. At a certain point, they appear to stop but this is the point where the ice melting from the glacier's snout is equal to the amount of new ice being pushed in from behind. The ice acts like a huge conveyor belt, bringing rock debris to the front of the glacier and depositing it there as moraine. Changes in the amount of ice delivered to the snout or changes in temperature will make the glaciers advance or retreat. Glaciers have been in retreat for more than 10,000 years.

V-SHAPED VERSUS U-SHAPED VALLEYS

Downcutting by water creates V-shaped valleys. On its journey to the sea, water snakes around, avoiding obstacles and following joints or other lines of weakness. Solid glacial ice, however, heads downhill in a straight line, avoiding nothing. Resistant rock, ridges or spurs are truncated by the advancing glacier, creating a smooth, polished U-shaped valley profile.

A sunbaker on a rock in Central Park, New York, is oblivious to the fact that its warm, comfortable surface was ground flat and scoured by ice and rock. This occurred as a glacier pushed southward during the last glacial advance 20,000 years ago.

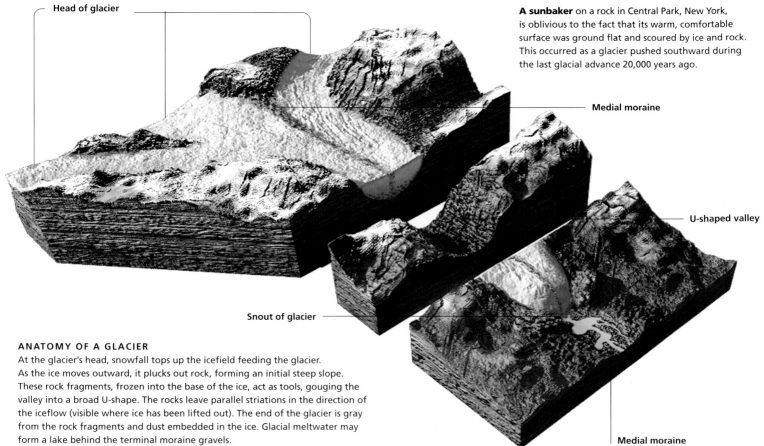

Head of glacier

Medial moraine

U-shaped valley

Snout of glacier

Medial moraine

ANATOMY OF A GLACIER

At the glacier's head, snowfall tops up the icefield feeding the glacier. As the ice moves outward, it plucks out rock, forming an initial steep slope. These rock fragments, frozen into the base of the ice, act as tools, gouging the valley into a broad U-shape. The rocks leave parallel striations in the direction of the iceflow (visible where ice has been lifted out). The end of the glacier is gray from the rock fragments and dust embedded in the ice. Glacial meltwater may form a lake behind the terminal moraine gravels.

↑ **Glaciers feature on Mt Blanc**, France, the highest
mountain in Europe at 15,719 feet (4807 m). Over
time, the permanent snow pack accumulating in
a number of cirques (circular-shaped depressions)
around the summit compresses into ice and moves
downslope under gravity. The smaller glaciers
coalesce to form larger ones that carve out
characteristic U-shaped valleys as they bulldoze
down the mountainside, pushing a pile of rocks and
dirt, known as terminal moraine, in front of them.

→ **This baby woolly mammoth**, *Mammuthus
primigenius*, affectionately known as Dima,
was found in 1977, well preserved in the frozen
permafrost of eastern Siberia, Russia. It is thought
that the six- to 12-month-old male died after
becoming stuck in mud about 39,000 years ago.
Freezing in ice is the the ideal way for fossils to
be preserved for study, as ice retains soft tissue.

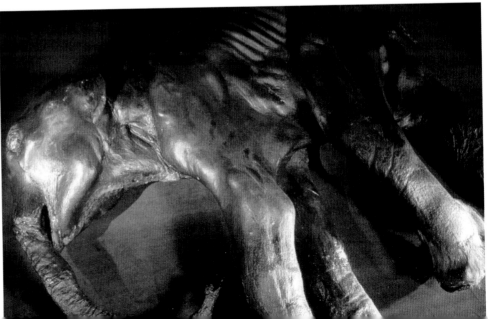

Rocks from space

Meteorites are thought to be the exploded remains of a former planet within our solar system. This theory accounts for the three types: stony meteorites from the planet's mantle; irons from the former core; and stony irons from somewhere in between. The pallasite is a rare but beautiful form of stony iron meteorite containing about equal amounts of iron and silicates with large gem-quality pieces of olivine. Meteorites and tektites are sought by collectors; the best sources are around the edges of the polar icefields. As glacial ice moves outward from its accumulation zone, it carries anything falling on it to the ablation or melting zone. Thousands of meteorites have been found there. Stony meteorites are the most common. A large impact has not occurred in written history but satellite images provide evidence of some huge ones that occurred in the past. The recently discovered Chicxulub crater hidden beneath jungle in Mexico's Yucatan Peninsula has a diameter of 112 miles (180 km) and was possibly responsible for the extinction of the dinosaurs. Vredefort, South Africa, the biggest and oldest known, is estimated to be 186 miles (300 km) across.

↑ **This backlit slice** of a pallasite meteorite reveals the bright green crystals of olivine within a shiny nickel-iron mass.

↓ **Chad's Aorounga impact crater** has a diameter of 10.5 miles (17 km). Buried under the sands of the Sahara, it shows up only in radar images.

Button-shaped or flanged disk tektite

Rod tektite becoming a dumbbell tektite

Dumbbell tektite becoming a teardrop tektite

TEKTITES

Tektites result from a comet or meteorite striking Earth. Molten fragments of rock are blasted high into the atmosphere, snap-cooling into a glass. They then fall back to Earth in a splash field, a zone where they are abundant. This may cross several countries. Tektites are normally dark in hue. However, moldavites are a very beautiful transparent green variety from Moldovia that can be faceted as gemstones. The diagrams show how the different tektite shapes can develop as a result of the molten fragments either flying or spinning through the air. Tektites have an etched surface and range 1–4 inches (2.5–10 cm).

TORINO SCALE OF METEOR EVENTS

Number	Possibility of impact
0	Zero or minimal chance of impact.
1	Impact extremely unlikely.
2	Close but not unusual encounter. Collision likely.
3	Close encounter, 1 percent or more chance of collision causing local damage.
4	Close encounter, 1 percent or more chance of collision causing regional destruction.
5	Close encounter, significant threat of collision causing regional destruction.
6	Close encounter, significant threat of collision causing global catastrophe.
7	Close encounter, extremely significant threat of collision causing global catastrophe.
8	Collision able to cause localized destruction. Such events occur every 50 to 1000 years.
9	Collision able to cause regional devastation. Likely to occur every 1000 to 100,000 years.
10	Collision able to cause global devastation and major climate change. Likely to occur less than every 100,000 years.

↑ **The Torino scale** is used to warn of the likelihood of meteorites hitting Earth. While small meteors are fairly common, the level of impact severity rises to 10, which indicates certainty of a mass extinction, such as that of the dinosaurs.

← **This meteor crater** near Flagstaff, Arizona, USA, formed when an object 100–165 feet (30–50 m) in diameter struck Earth about 50,000 years ago.

Rock structures

Many common rocks, such as sandstone, granite, slate, limestone and marble, have long been used for building and carving. Some projects have been on a vast scale, such as the Great Wall of China or the pyramids of Egypt. Historically, rocks only needed to be durable, free of fractures and readily available to be useful for such structures. In metamorphic rocks, such as slate, the planes of weakness are used to advantage as slate is easily split into sheets for roofing or paving. Today, bricks, steel, reinforced concrete and glass have largely replaced building stone, although it is still commonly used for decorative purposes when a ambience of luxury and elegance is required. Polishing will readily bring out a rock's internal structure and the beauty of its constituent minerals.

↑ **A hidden city** at Cappadocia in Turkey was dug into the soft volcanic ash by early Christians escaping persecution from the Romans. Some dwellings have tunnels and shafts that extend deep underground.

→ **The ancient tombs** of Petra, Jordan, lay hidden for 2000 years in a maze of narrow canyons called siqs. These are not the impressive buildings they appear to be, but are a series of facades carved into the sandstone canyon walls. The beautiful swirly yellow, mauve, orange and red bands resulted from rhythmic deposition of manganese and iron by groundwater flowing in the sandstone. The name Petra is from the Greek word for rock.

Stonehenge, in Wiltshire, England, was constructed in several stages, and by different peoples, between 3000 and 1500 BC. It is one of a number of circle sites in the area, the exact religious or ceremonial functions of which are not fully understood. The large pillars and lintels of the later, outer circle are a hard, siliceous sandstone known as sarsen, which was transported from Marlborough Downs about 20 miles (32 km) away. The inner bluestone circle was built earlier. It is mainly dolerite from the Preseli Hills in Wales.

↑ **The presidents' heads** at Mt Rushmore, USA is one of the world's most ambitious modern sculptures. It was carved into a granite mountain between 1927 and 1941 by John Gutzon Borglum and 400 workers.

Minerals

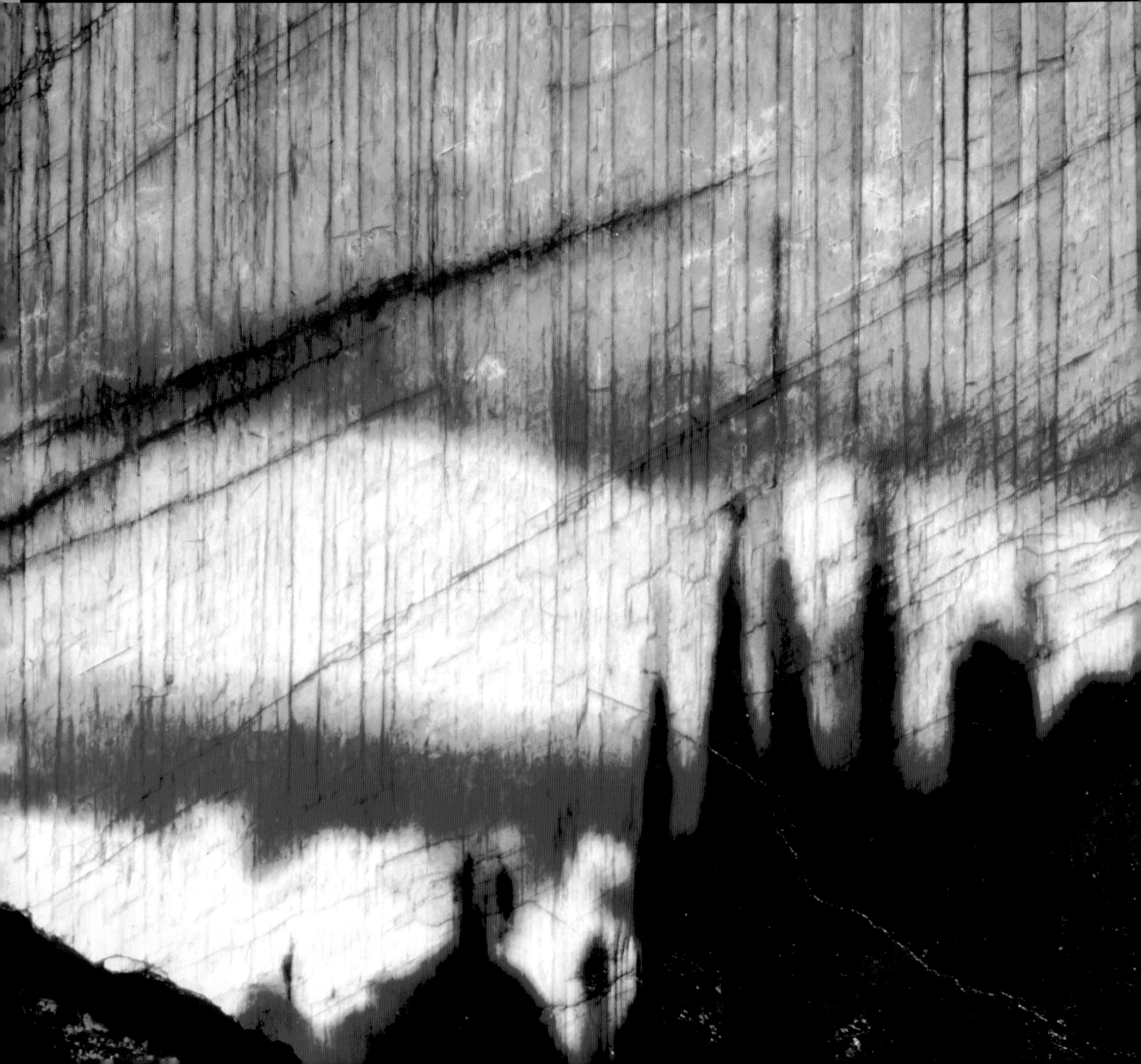

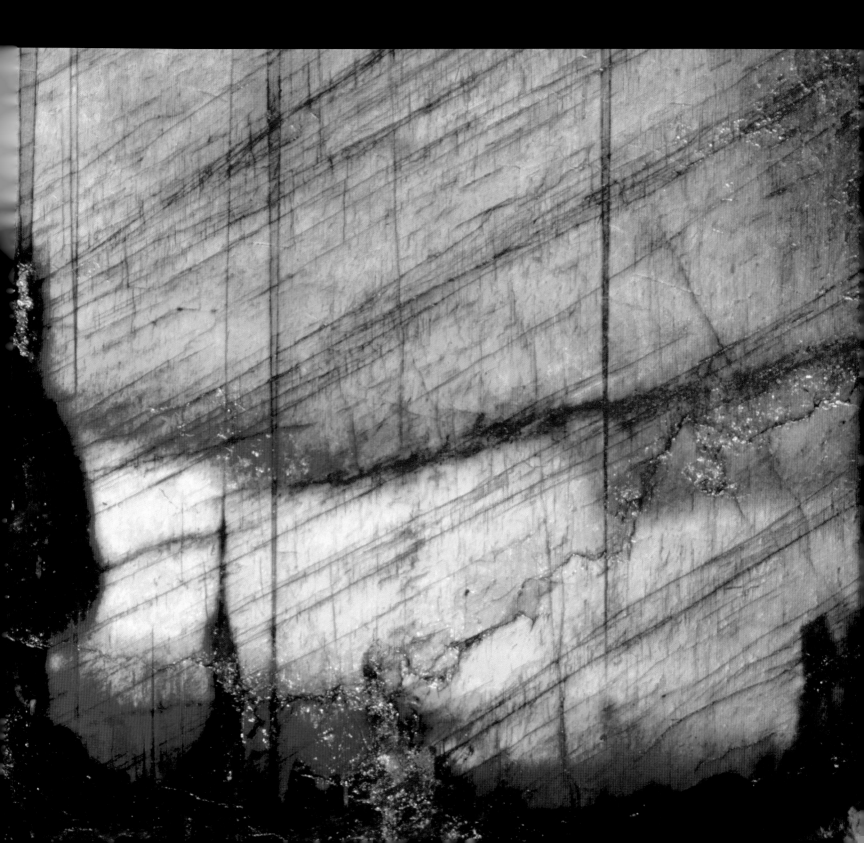

Previous page Labradorite, one of Earth's mineral
treasures, displays a blue flash when turned toward light.

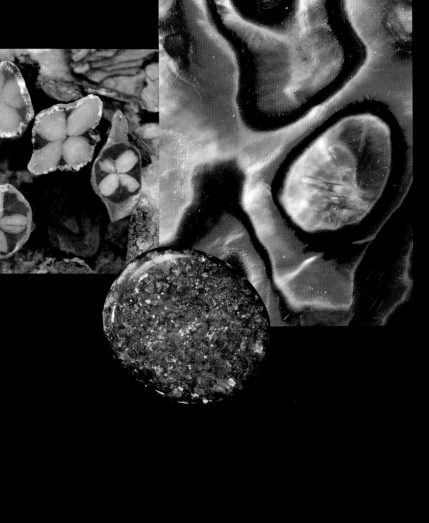

Thousands of minerals may be found on Earth's surface, each with unique composition, crystal structure and properties. Often properties such as hardness, color, luster, transparency and sheen combine to make a mineral or rock of exceptional beauty, which is highly sought after for use as a gemstone.

Understanding minerals

A mineral is a naturally occurring inorganic crystalline solid with a definite chemical composition. Minerals are the building blocks that make up rocks, although only a handful of these, known as the rock-forming minerals, are very common. People of ancient civilizations learned to carve and polish minerals for adornment and decoration, and discovered their useful physical properties. Minerals play an essential part in daily life—we extract the numerous elements we need from them to run our high-tech activities. Even the glasses we drink from and the ceramic plates we eat off are derived from minerals. Minerals are made up of combinations of Earth's 94 naturally occurring elements and they have distinct, measurable physical and chemical properties. Referring to these properties, geologists, armed with a variety of testing equipment, are able to identify any mineral specimen. In the gem industry, rapid, inexpensive but accurate testing is crucial because cut gemstones, identical to the naked eye, may vary several orders of magnitude in price.

→ **Multicolored tourmaline crystals** show why color cannot be relied on to identify certain gemstones. However, the triangular shape and vertical striations identify this mineral as tourmaline.

↓ **Shape, form and** color distinguish the four minerals in this pegmatite: the yellowish microcline; bladed books of silver muscovite; octahedral fluorite; and tabular pink apatite.

MOHS' SCALE OF HARDNESS

Talc	1
Gypsum	2
Calcite	3
Fluorite	4
Apatite	5
Orthoclase	6
Quartz	7
Topaz	8
Corundum	9
Diamond	10

OTHER PROPERTIES

Specific gravity is a mineral's density. It ranges from light, such as sulfur and graphite (specific gravity of 1–2); to very heavy minerals (over 6), with platinum and native gold being the heaviest (19).

Magnetism indicates a high iron content and that the mineral will attract a magnet, as do pyrrhotite and maghemite. Magnetite will attract iron to itself. The first compass was lodestone (magnetite).

The acid test will indicate the presence of some carbonate minerals (such as calcite) and carbonate rocks (such as limestone, marble and chalk).

Thermal conductivity is how well a mineral can conduct heat. Minerals that conduct heat well will feel cool. Hence, diamonds have the nickname ice. Poor conductors, such as amber and jet, will feel warm to touch.

HARDNESS

A mineral's hardness is determined by scratching it with an object or another mineral of known hardness. Mineralogist Friedrich Mohs (1773–1839) developed a set of reference minerals, ranging from hardness 1 (talc) to 10 (diamond). Some everyday objects can also be used as references: a fingernail (2.5), copper coin (3.5), glass (5.5) or a steel knife (6.5). Hardness is a destructive test as it leaves a scratch. Minerals below 7 are unsuitable for day-to-day wear as gemstones. A diamond's hardness ensures it can withstand a lifetime of wear.

Mohs' scale of hardness versus true hardness, displayed graphically, reveals just how hard diamond and corundum are. Corundum (ruby and sapphire) is significantly harder than topaz, and diamond is about three times as hard as corundum. Hardness makes diamond and corundum truly precious gems.

Color, sheen and luster

What a mineral looks like is often the most important clue in identifying it. Color, sheen and luster are the easiest properties to observe in a mineral. To best use this information when examining a specimen, hold the mineral up to the light and try to look through it, or through a thin edge if it is too dark. Surprisingly, the attractive minerals labradorite and opal appear an unpleasant, pale and smoky yellowish brown. This is their true color. The other spectacular optical effects they display are their sheen. Sheen is the result of light interference from structures or tiny inclusions within the stone, while luster or shine is due to light reflecting off the surface of the stone. Sheen and luster can be observed properly only on non-water-worn surfaces or crystal faces, or in polished gemstones. Streak is the color of the powdered mineral and it can vary from the mineral's color. Rubbing a mineral on an unglazed white porcelain tile shows its streak. Silvery gray hematite displays a cherry-red streak, but most multicolored minerals, such as zircon, have a gray–white streak.

→ **Labradorite is a** variety of feldspar that displays a distinctive pearly blue iridescent sheen known as schiller.

↓ **These spectacular columns** of watermelon tourmaline display a color difference from red to green along the length of the crystal. This is a result of slight changes in the composition of the mother liquor while the columns were growing.

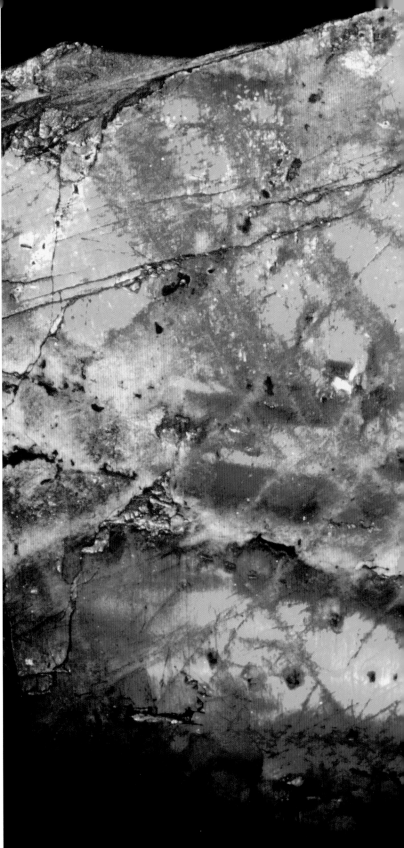

COLOR

Minerals are colored by the elements that make up their chemical structure. Malachite is always green, turquoise is always blue and sulfur is yellow. Color is a diagnostic property for these self-colored or idiochromatic minerals. Other minerals are perfectly colorless when pure, but are usually colored by trace amounts of impurities.

A good example is beryl, which is practically worthless when pure. But, when colored green by trace amounts of chromium, it becomes emerald, a very valuable gemstone. Some of these allochromatic minerals, such as tourmaline, quartz, zircon and fluorite, can occur in almost every imaginable color.

Rhodonite is always pink, being colored by the major element component, manganese.

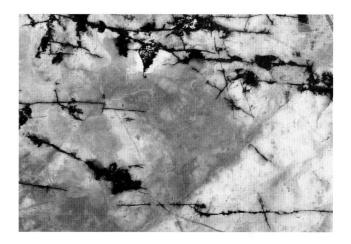

SHEEN

Some minerals display effects caused by the light reflecting off the mineral's internal structure or inclusions. Chatoyancy occurs where a bright beam runs across the stone, while asterism is chatoyancy displayed in more than one direction, creating a star of bright light. This can be seen in star diopside (two directions) and star ruby or sapphire (three directions). Play-of-color occurs in precious opal where light rays are refracted by layers of microscopic silica spheres, creating vivid patches of color. Adularescence is the floating blue cloud effect seen in moonstone and some transparent opal, while iridescence is the rainbow effect seen on the inside of pearly shells as well as in obsidian and labradorite.

Cat's eye chrysoberyl contains fine, oriented needle-like fibers and displays a single chatoyant beam.

LUSTER

Luster refers to the appearance of a mineral's surface and it depends on how light is reflected. It is a purely surface-related effect. The superb adamantine luster of diamond and the sub-adamantine luster of zircon are distinctive and make these gemstones appealing. Most minerals, such as quartz, beryl and topaz, are vitreous or glassy. Other lusters include resinous (amber, jet), pearly (talc, pearl), waxy (turquoise), silky (asbestos, tigereye) and metallic (hematite, galena). If a mineral displays no luster it is termed dull or earthy (kaolinite).

The bright metallic luster of this pyrite from Peru is typical of many of the metallic minerals.

Optical effects

Many optical properties that affect color can help identify a mineral. Pleochroic minerals, such as sapphire and tourmaline, change color when viewed in different orientations. Some do so spectacularly—cordierite (or iolite) is a deep blue gemstone but when viewed down the long axis of the crystal it appears practically colorless. Color-change minerals will change color, sometimes completely, under different lighting conditions. Fluorescent minerals luminesce under ultraviolet light, with fluorite being a classic example. Phosphorescent minerals, such as opal, continue to glow even after the ultraviolet light is switched off. Dispersive or fiery minerals, such as diamond, separate white light into scintillating rainbow flashes. When quartz pieces are rubbed together, they glow with a cold blue light. This property, called triboluminescence, was known to the ancient Incas. Images double up when viewed through some minerals, such as calcite, tourmaline and zircon. As diamond does not show doubling, there is no difficulty in separating it from an otherwise indistinguishable cut zircon.

COLOR CHANGE

Every mineral changes color slightly when viewed under different lighting conditions. For this reason gemstone graders and valuers work in white-painted rooms with special standardized lighting. Only very few minerals change color dramatically enough to be noticeable. These include sapphires, some garnets and alexandrite, a particularly rare variety of chrysoberyl, named for the Russian Czar Alexander. The spectrum of alexandrite is so delicately balanced that when viewed in the blue–green-dominant light of the Sun it will impart this hue. However, at night, under the red-dominant light of an incandescent bulb, a red hue will dominate. This feature makes it a most valuable gem.

↓ **Alexandrite displays** the most spectacular color change of all minerals. The same gemstone is red in artificial light, but in daylight it appears green.

→ **Fluorescent minerals** glow in the dark under invisible ultraviolet light. They frequently do so in spectacularly bright colors that are often quite different from their daylight colors.

SPECTRUM

When white light passes through a mineral, parts of its visible color spectrum may be removed by the crystal lattice. The remaining spectrum can be observed using a spectroscope and a white light source—the missing parts appear as distinct black lines. Different minerals may have diagnostic spectra.

Strong dispersion is seen in the flashes of rainbow color in a diamond. White light entering a transparent gemstone is broken down into its spectral colors as each is bent at a slightly different angle. The greater the difference between the red and the blue rays, the stronger the dispersion will be.

High dispersion — White light — Diamond

Low dispersion — White light — Quartz

REFRACTION

Refraction is a measure of how far a ray of light is bent when passing into a transparent mineral. Some minerals have up to three values of refractive index depending on the direction the light beam is traveling when it goes through the crystal lattice. The difference between the highest and lowest refractive index within a mineral is called its birefringence. If this value is high enough, image doubling can be seen with the naked eye.

→ **Iceland spar calcite** exhibits doubling or double refraction. Light reflecting off the needle and passing through the crystal is polarized (split up) and refracted (bent). The thicker the piece of calcite, the farther apart the images appear.

Clarity and inclusions

Looking into the inner world of a mineral with a hand lens or a microscope is one way to identify it. A mineral's clarity is an important property. Certain minerals, such as topaz, are almost always transparent while others, such as malachite, are never transparent. Some minerals contain inclusions identifiable by eye that are diagnostic. Sapphire and ruby often contain fine, hairlike rutile needles, called silk, oriented in three directions. Olivine commonly contains eight-sided chromite crystals surrounded by disk-shaped tension halos that look like lily pads. Demantoid garnets also have chromite with bunches of radiating fibrous byssolite crystals that resemble a horse's tail. Emerald is often filled with a multitude of spiky-shaped cavities that contain fluid and gas bubbles called jardin. Scientific study of inclusions within a mineral allows us to ascertain formation depth, temperature of crystallization and even its age. The surface of the mineral is carefully polished away until the inclusion is exposed. It can then be identified using an electron microprobe.

⇢ **The crystal inclusions and fingerprints** in this sapphire were the result of a guest crystal being included while the sapphire was growing. Changes in size during cooling created this tension fracture or fingerprint around the guest crystal.

→ **These droplets in amethyst** tell an interesting story. As a crystal grows in the mother liquor, it may preserve droplets of this liquid within itself.

↓ **As this ruby cooled**, tiny böhmite needles evolved out in three directions at 120° angles to one another. Each needle reflects light and if there are enough needles, a six-pointed star will be seen in the polished ruby.

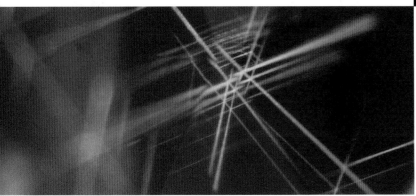

INCLUSIONS

Minerals often contain inclusions. These may be older minerals that existed for quite some time before the host crystal grew around them. Such inclusions are abundant in metamorphic minerals that grow in solid rocks as the inclusions cannot move out of the way. Other minerals may start growing at the same time but one that grows more quickly may include the others. These inclusions can be analyzed and are useful for revealing the conditions in which the host crystal grew. Droplets of fluid in which the crystal is growing may also be included at this time. Even after the host crystal has fully grown, further inclusions may develop. These include tension cracks or fingerprints around other inclusions, minerals that evolve within the host as the temperature drops, and even halos of radiation damage around radioactive minerals. Some gemstones rely on their inclusions to make them attractive, such as the tiny mica flakes in sunstone.

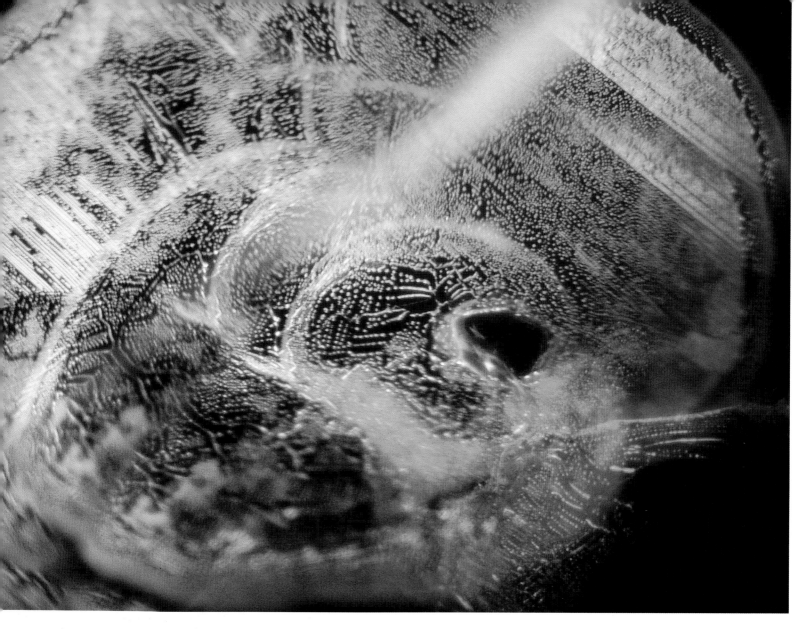

TRANSPARENCY

Transparency is the ease with which objects can be
seen through a mineral. Many minerals are transparent
when pure, such as quartz, sapphire and tourmaline.
Minute inclusions, however, can reduce a mineral's
transparency and reduce its value as a gemstone.
Semi-transparent minerals, such as varieties of
moonstone, allow light to pass, making an object
behind still visible but blurred. Translucent minerals,
such as chalcedony and nephrite, allow some light to
pass through but no object behind would be visible.
Opaque minerals transmit no light and include most
of the metal sulfides and oxides.

Transparent
such as quartz

Semi-transparent
such as some moonstones

Translucent
such as chrysoprase

Opaque
such as malachite

Crystal systems

The shape that a crystallizing mineral will take reflects the internal arrangement of its atoms and molecules. Perfect crystals treasured by collectors are rare in nature, but even a distorted crystal or broken piece may be enough to measure the angle between the faces in order to classify it into one of the six crystal systems. The systems range from cubic, which has the most elements of symmetry, to triclinic, which has the least symmetry. There is no easy rule but many of the elemental minerals, such as native copper, gold, silver, platinum, arsenic and diamond, belong to the cubic system. This is because all their atoms are the same, enabling them to pack into the minimum space configuration. The other elemental minerals—sulfur, graphite and bismuth—are packed close into a hexagonal arrangement and therefore belong to the hexagonal system. Some minerals have identical composition but belong to entirely different systems because they have formed under quite different conditions. Carbon in the form of diamond crystallizes in the cubic system whereas carbon as graphite is hexagonal. Such minerals are known as polymorphs.

→ **Striated, trigonal crystals** of deep green tourmaline display a columnar habit. They are intergrown with a transparent, trigonal, columnar crystal of quartz. This specimen comes from a mine in Itinga, Brazil.

↓ **This hexagonal crystal** of violet–purple apatite displays a tabular habit. It is intergrown with a fine matrix of quartz, feldspar and mica. It comes from Kumar province in Afghanistan.

An orange equant crystal of spessartite garnet, from Ramona in California, USA, is intergrown with a trigonal prism of black tourmaline and fine, bladed crystals of albite feldspar.

This amazonite feldspar formed stubby triclinic crystals together with prismatic smoky quartz in a pegmatite vein from Colorado, USA.

CRYSTAL SYSTEMS

The six major crystal systems into which all minerals crystallize are shown here, together with a number of perfect crystal forms as examples. The systems, from the most symmetrical to the least are: cubic, three axes of equal length at right angles (90°) to one another; tetragonal, three axes at 90° to one another, two of equal length with the third being either longer or shorter; hexagonal (including trigonal), three axes of equal length at 120° to one another with the fourth at 90° to the rest being longer or shorter; orthorhombic, three axes of unequal length at 90° to one another; monoclinic, three axes of unequal length, two at 90° with the third being greater than 90°; and triclinic, three axes of unequal length and none at right angles, being the least symmetric. The crystal system is a particularly important factor in determining the habit or form a mineral may take.

Crystal system   **Mineral examples**

Cubic   galena   magnetite   pyrite   almandine   tetrahedrite

Tetragonal   rutile   zircon   chalcopyrite   scapolite   apophyllite

Orthorhombic   sulfur   cerussite   olivine   enstatite   barite

Monoclinic   orthoclase   gypsum   realgar   augite   wolframite

Triclinic   albite   kyanite   chalcanthite   sassolite   rhodonite

Hexagonal and trigonal   beryl   apatite   tourmaline   quartz   calcite

Habit

Habit is the shape in which a mineral grows. Many minerals prefer to grow in a typical habit, such as the yellow-gold cubic crystals of pyrite. Descriptive terms for habit refer to single crystals and aggregates. A crystal's size and the shape of its faces are determined by the mineral's atomic structure; that is, how quickly it prefers to add atoms (or grow) in certain directions. Minerals growing equally in all directions will produce equant crystals. Faster growth in one direction will produce columns or even needles. Much slower growth in one direction will produce tabular crystals or blades. While some crystals can grow into cavities left by gas bubbles or into the center of veins, most do not get the opportunity to grow into open space. Such crystals often intergrow with one another and are forced to form crystalline aggregates.

CRYSTAL HABITS

Single crystals display a variety of habits. They can be equant or blocky (microcline, cassiterite); columnar (tourmaline, beryl, quartz); acicular or needle-like (natrolite, millerite); barrel-shaped (sapphire); bladed (kyanite) and tabular or flattened (ruby, barite). More descriptive terms can aid identification if these features are visible. Members of the cubic system may be described as forming cubes, octahedra or dodecahedra. Columnar crystals may have three, four, six or 12 faces. They may have a flat top known as a pinacoid or come to a point called a pyramid. Crystal face detail is also diagnostic: they can be flat, smooth, striated or etched.

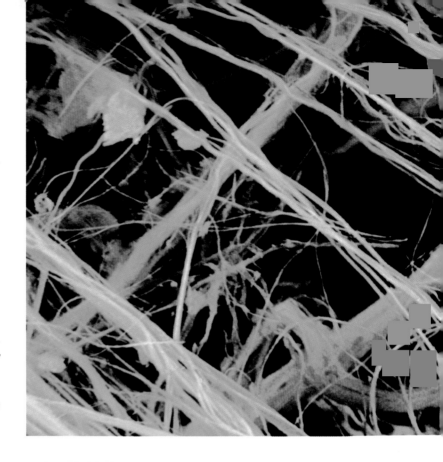

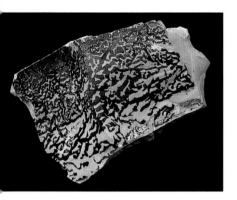

↑ **A scanning electron micrograph** reveals the fibrous habit of asbestos. These fireproof fibers make an excellent insulator but unfortunately they are hazardous to human health.

←← **The dendritic habit** of iron oxide produces a dark mosslike pseudofossil along the joint planes in chert. This specimen is from Newcastle in New South Wales, Australia.

← **Transparent columnar** prisms of clear quartz are intergrown with bladed crystals of golden yellow star muscovite. This example comes from Minas Gerais, near Aracuai, Brazil.

HABITS

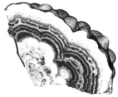
Botryoidal

Pisolitic

Columnar

Tabular

Acicular

AGGREGATE HABITS

Aggregates are a mass of interlocking crystals that form as a result of close-knit growth of one or more mineral species. This may result from growth in a confined space or be the crystal's preferred habit. Aggregate types have descriptive names. Granular is seen in the common interlocking texture of calcite in marble or of quartz in granite. Fibrous is visible in parallel bunches of long, thin crystals like asbestos. Micaceous, foliated or lamellar aggregates occur where minerals intergrow as thin, easily split plates (mica, graphite, molybdenite). Stellate crystals radiate outward from a central point (natrolite, gypsum). Dendritic, arborescent, mosslike or treelike crystals form highly complex patterns that often resemble fossils (manganese oxide in moss agate, native copper, silver and gold). Reticular aggregates are oriented crystals that form a reticulated network (rutile, cerussite). Rosettes have a flowerlike pattern, as displayed by desert rose gypsum or iron rose hematite. Interesting colloform habits result from the non-crystalline or microcrystalline precipitation of minerals. These include: botryoidal, or grapelike (malachite); reniform, or kidneylike (hematite), globular (wavelite); stalactitic, as found in cave formations (calcite, chalcedony); and pisolitic or oolitic, which are concentric spheres that grow around a nucleus (bauxite, laterite).

↑ **This malachite nodule** from Musoni, Congo, displays the typical botryoidal habit, which resembles a bunch of grapes. Malachite is an ore of copper and this attractive rich green mineral is often used as a gemstone. When cut through and polished it shows a series of attractive light and dark green concentric growth rings that run parallel to the outside surface.

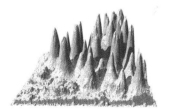

Stalactitic

Reniform

Stellate

Dendritic

Equant

Cleavage and fracture

Strike a mineral with a hammer and it will cleave, or fracture, into pieces. The appearance and shape of these pieces can be quite diagnostic in identifying the mineral, although this test is unsuitable for fine specimens. Cleavage is the term used for breakage along definite planar surfaces. These may look like crystal faces but should not be confused with them. Cleavage occurs along planes in the crystal where the strength of the bonding between the atoms is weakest. The classic example is in mica where the so-called van der Waals bonds between the silicate sheets are extremely long and weak, allowing them to come apart easily. On the other hand, if the bonds are of equal strength in all directions, such as in quartz, then the mineral will not cleave. Cleavage faces are described as being perfect or eminent if they are straight, smooth planes (mica, calcite, diamond); good (fluorite, pyroxene); or imperfect or poor if they are uneven (beryl, zircon). Cleavage planes are referenced to the crystal faces that they parallel and are also described as either being easy to achieve, such as in mica and fluorite, or difficult, such as in diamond. In Amsterdam, the center of the world's diamond industry, cleavers use this property to shape and fashion this hardest-known mineral.

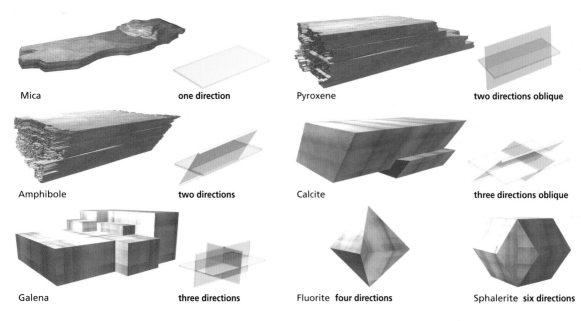

Mica   **one direction**

Amphibole   **two directions**

Galena   **three directions**

Pyroxene   **two directions oblique**

Calcite   **three directions oblique**

Fluorite **four directions**

Sphalerite **six directions**

CLEAVAGE DIRECTION AND ORIENTATION

The direction and number of cleavages are described as follows: one direction (as found, for example, in mica); two directions, with the planes at right-angles to one another (as in pyroxene and orthoclase); two directions, with the planes intersecting at an oblique angle (as in amphibole); three directions at right angles, forming cubes (as in galena), three directions at oblique angles, forming rhombs (as in calcite); four directions, forming octahedra (as in diamond and fluorite); and six directions forming rhombic dodecahedra (as found in sphalerite).

FRACTURE AND TENACITY

Minerals that do not cleave easily tend to fracture irregularly. The appearance of this fracture surface can be diagnostic. Most minerals display a conchoidal (shell-like) fracture, which looks like a series of curved, concentric undulations spreading out from the point of impact. This fracture type was well understood by prehistoric peoples working obsidian and flint to make stone tools. Tough minerals, such as jadeite and nephrite, tend to have a hackly fracture like that of broken cast iron. Kaolinite and chalk break with an earthy fracture. Tenacity is a mineral's resistance to breaking. Most minerals, including pyrite ("fools gold"), will either cleave or fracture but only a few display diagnostic tenacity. Gold and other native metals are ductile or malleable. The micas are elastic, so thin cleavage sheets can be rolled up and will spring back to their original shape when released. Molybdenite is flexible—it bends without breaking but it will not spring back.

↑ **A close-up view of** a crystal of muscovite shows the one perfect cleavage of this form of mica. Using just a fingernail, the thin, hexagonal plates seen here can be lifted quite easily, peeled off and even rolled up, as they are extremely flexible.

↗ **This Paleolithic axe** from Surrey in England is made from flint. Flint has no cleavage but it fractures randomly when hit. It displays a typical smooth, shell-like conchoidal fracture.

→ **Two directions** of incipient cleavage are visible as parallel cracks in this piece of gypsum. If dropped onto a hard surface, the crystal would break along these planes.

Identifying minerals

Minerals have distinct, measurable physical and chemical properties by which they can be grouped and identified. There are now roughly 4000 mineral species known to science, with more discoveries being made. Minerals are commonly broken down into eight classes according to their chemical composition, based on the original 1837 work by pioneering mineralogist James Dana. His *System of Mineralogy* was continued by his son Edward and later developed by others. The first class contains elements, such as copper, gold, silver, carbon and sulfur, which can occur alone in their natural state. These minerals are known as native elements. The remaining classes contain minerals that are combinations of various atoms and are named for their non-metallic component. These are sulfides such as chalcopyrite, a copper sulfide; oxides such as cuprite, a copper oxide; carbonates such as malachite, a copper carbonate; halides such as atacamite, a copper chloride; phosphates such as turquoise, a copper phosphate; sulfates such as chalcanthite, a copper sulfate; and silicates, the rock-forming minerals. A ninth class is sometimes added for organic "minerals" such as amber, jet and pearl.

→ **This green columnar tourmaline** is in a matrix of bladed albite, deep purple lepidolite mica and quartz. This specimen is from the Pederneira Mine near São José de Safira, Minas Gerais, Brazil.

↓ **Well-formed, blocky, white microcline feldspar** crystals are partially covered with greenish yellow plates of muscovite mica. This specimen was mined from a pegmatite at Galileia, Minas Gerais, Brazil.

POLYMORPHISM

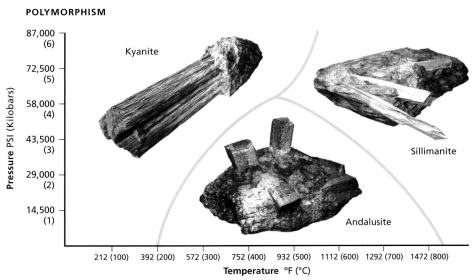

Pressure PSI (Kilobars) / Temperature °F (°C)

Kyanite

Sillimanite

Andalusite

← **Pressure–temperature** in the formation of kyanite, andalusite and sillimanite is illustrated here. These minerals are identical in their composition. Their presence in rocks tells us the conditions under which they formed. Andalusite indicates low pressure and high temperature; kyanite indicates higher pressure and low temperature; sillimanite indicates high pressure and temperature.

SAME COMPOSITION, DIFFERENT MINERAL

Polymorphism occurs when a single element or compound can exist as two or more distinct minerals. That is, it can grow as different crystal structures depending on the environment (temperature and pressure conditions) in which it finds itself. Polymorphism is quite common—examples include graphite and diamond; aragonite and calcite; and andalusite, kyanite and sillimanite.

SAME STRUCTURE, DIFFERENT MINERAL

Isomorphism occurs when an identical structure is shared by two different minerals. A solid solution exists when an atom is able to substitute for another atom of similar charge and size in a mineral. The spinel group, with its 22 member minerals (including gem spinel, magnetite and chromite), is the best example. Metals such as magnesium, zinc, iron, chromium and aluminum can interchange freely while the crystal is growing. Likewise, the tourmaline group also has several members—elbaite, dravite, schorl, buergerite, liddicoatite, povondraite and uvite—as there is substitution between elements.

SAME FORM, DIFFERENT MINERAL

Pseudomorphism occurs when there is a complete change of composition through chemical replacement. The new mineral mimics the crystal system and habit of the old but it may be a completely different color, such as malachite (green) replacing azurite (blue).

↑ **Diamond and graphite** are made from the same element. This single, octahedral, gem-quality diamond crystal is shown with two rounded crystalline aggregates of industrial-quality diamonds on a background of graphite. Both these minerals are physically different in every respect, although they are chemically identical, being made entirely of carbon.

Native elements

Minerals in this class are made up of a single type of atom and are referred to as the elemental minerals. They include a number of metals, such as copper, gold, silver, platinum, nickel-iron, antimony, bismuth, arsenic and mercury. It was the discovery that shiny lumps of gold and copper could be beaten and shaped without breaking that launched civilization from the stone age into the metal ages. Today, apart from gold, the native metals are of relatively minor importance compared with the metal sulfides and oxides, which are mined and processed at very low grades and are our source of metals. Important non-metallic elemental minerals include sulfur, graphite and diamond.

This specimen of crystalline native gold on quartz has a beautiful arborescent (treelike) habit and a sparkly metallic luster. Gold is soft (2.5–3), very heavy, malleable and melts at 1942°F (1061°C). It is these properties that distinguish it easily from pyrite or "fool's gold." Gold will not tarnish and is an excellent conductor of both heat and electricity.

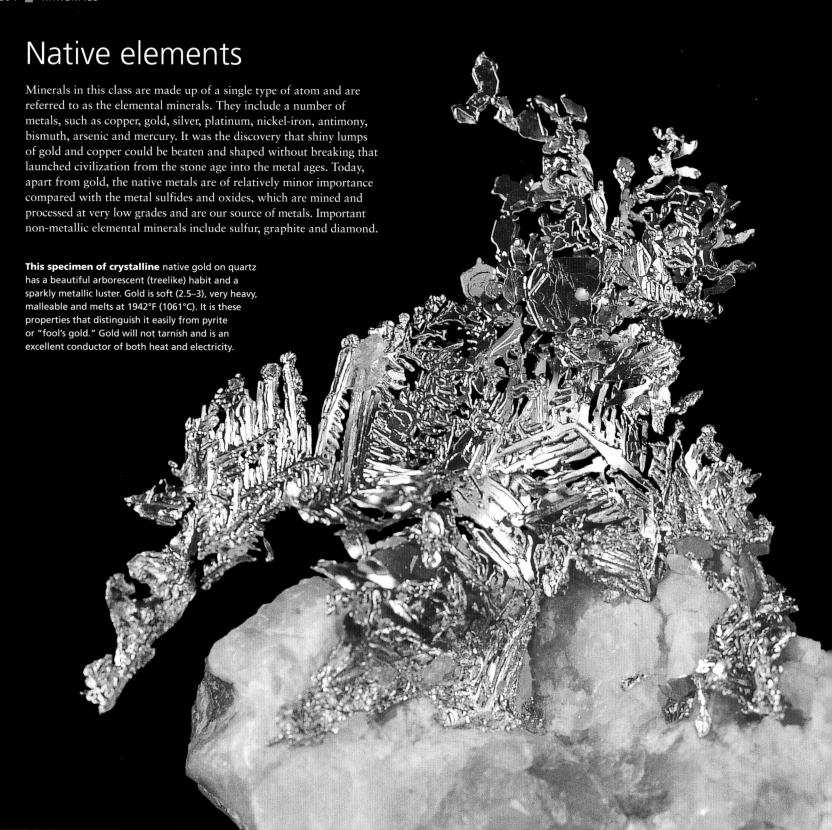

COPPER, GOLD AND SILVER

Discovered at least 8000 years ago, gold was the first of the native metals to be worked, later followed by copper and silver. Strong and pliable, these metals could be made into tools, weapons and ornaments. About 6000 years ago another important discovery was made, when a small amount of tin was added to molten copper. This resulted in the alloy bronze, which was found to be harder and more useful than either of its two constituents. This discovery occurred in Mesopotamia, probably as a consequence of the proximity of both copper and tin mines. The Sumerians were the first to use bronze in commerce. Copper has retained its role—its abundance, ductility and conductivity have ensured its value in electrical wires and cables. Silver and gold remain important in the jewelry industry, and gold has long been an international monetary standard.

THE GOLD RUSHES

A gold rush is a stampede of prospectors, adventurers, merchants and others to a newly discovered gold field, all hoping to make their fortune. The discovery of gold in California in 1848 lured some 40,000 people within the space of two years, most of whom never struck it rich. It did, however, stimulate economic growth in commerce, transport, agriculture and industry in the region. Later important gold rushes took place at various locations in Australia (1851, 1893), at Witwatersrand in South Africa (1886), and the Klondike in Canada (1896).

↗ **These native copper** crystals display branching or arborescent habit. Copper, red when fresh, may have a greenish weathering film of malachite. It is soft (2.5–3), heavy, malleable, conducts electricity and melts at 1980°F (1082°C).

→ **Gold rushes** have tempted thousands of fortune hunters, many anxious to escape their lackluster routine lives. The lure of the gold fields has disrupted societies and made or broken many lives.

Canada, 1896
California, USA, 1848
Venezuela, 1986
Brazil, 1980
Philippines, 1983
Papua New Guinea, 1988
South Africa, 1886
Australia, 1893
Australia, 1851

Native mercury droplets. This liquid metal solidifies at –38°F (–39°C), is an excellent conductor of electricity and is very heavy.

This native silver shows typical branching or arborescent habit. The crystals are soft (2.5–3), heavy, malleable and melt at 1,760°F (960°C).

A native sulfur cluster made up of orthorhombic crystals displays blocky habit. Sulfur is light, soft (1.5–2.5), fragile, melts at 246°F (119°C) and is always a brilliant yellow.

Metallic minerals

The metallic minerals are the principal sources of all the metals we use in our daily lives. The chance realization that metals could be extracted from certain heavy, colorful rocks placed in a hot campfire or kiln led to the discovery of lead and mercury, as well as new sources of copper and silver. The last metal to revolutionize the lives of ancient peoples was iron. Although more difficult to extract than the earlier metals, needing a hotter-burning coal fire, its abundance and strength changed the face of civilization. By 1100 BC, iron was in widespread use, and tools and weapons were vastly superior to those of the preceding Bronze Age. Improved technology continued to give access to a range of new metals. Many metallic minerals, such as hematite, pyrite and chrysocolla, are quite attractive and some are cut and polished for use in jewelry.

↑ **Lustrous, brassy chalcopyrite** crystals sit alongside clear crystalline quartz and calcite. This specimen is from the Cavnic district in Romania.

← **These three shiny**, cubic crystals of galena are perched on a splinter of white siliceous dolomite. They have octahedral faces.

→ **These rhombohedral** twinned crystals of cinnabar (mercury sulfide) are in a matrix of transparent, fine crystalline quartz, from Fenghuang Hunan, China.

PRINCIPAL ORE MINERALS

The term ore refers to those minerals from which valued metals can be economically extracted. Although most metals can be derived from several different minerals, the most important ores are listed in the table (*right*). These are all sulfides or oxides. Aluminum, although abundant, was impossible to extract on a large scale before the 1870s because of the vast amount of power required to release it from its ore, bauxite. Other important modern metals include titanium, chromium, tungsten and platinum.

PRINCIPAL MINERALS USED FOR THE EXTRACTION OF METALS

Ore mineral	Composition	Metal extracted
Hematite	Fe_2O_3	iron
Magnetite	Fe_3O_4	iron
Chalcopyrite	$CuFeS_2$	copper
Sphalerite	ZnS	zinc
Argentite	Ag_2S	silver
Galena	PbS	lead
Cinnabar	HgS	mercury
Stibnite	Sb_2S_3	antimony
Arsenopyrite	$FeAsS$	arsenic
Molybdenite	MoS_2	molybdenum
Rutile	TiO_2	titanium
Cassiterite	SnO_2	tin
Pyrolusite	MnO_2	manganese
Uraninite	UO_2	uranium

← **Hematite** (iron oxide) is the most important ore of iron. It shows classic reniform or kidneylike habit.

← **These vivid yellow,** square, tabular plates are of wulfenite (lead molybdate). It occurs in the oxidized zone of lead deposits.

← **These blocky, monoclinic** crystals are of blood-red realgar (arsenic sulfide). This mineral is used in the manufacture of paint and fireworks.

↓ **These well-formed** octahedral crystals are of cuprite (copper oxide) formed on chrysocolla (copper silicate). Both minerals are ores of copper.

Evaporites

The evaporites are minerals that are, to varying degrees, soluble in water. These minerals include halides (halite, sylvite, carnallite), nitrates (niter), borates (ulexite, colemanite), sulfates (gypsum, glauberite) and carbonates (calcite, dolomite, aragonite). As lakes dry out and the chemical concentration increases, these minerals precipitate, sometimes as spectacular crystals. Walking on one of these dry salt lakes, the zones of different minerals can be observed, with the first to crystallize being the least soluble. Halite is extremely soluble and will precipitate only when the water is nearly gone. Huge evaporite deposits occur in newly formed continental rifts when seawater pours into the opening end and starts to evaporate. The Mediterranean Sea was a vast evaporite basin during the Miocene (23 to 5 mya).

→ **Halite or rock salt** is the most common evaporite and the last to precipitate. Evidence of its high solubility can be seen in the form of rainwater channels etching this specimen.

↓ **This dry salt lake** in Badwater salt flats, Death Valley, USA, is filled with residual evaporite minerals.

↑ **This delicate cluster** of aragonite (calcium carbonate) crystals shows a radiating fibrous habit. Aragonite is semi-hard (3.5–4) and fragile. As it readily dissolves in cold, dilute hydrochloric acid, applying this acid can be diagnostic. In nature, aragonite is one of the first minerals to crystallize from solution in a salt lake as the water evaporates.

↑ **These gypsum** (calcium sulfate) crystals are growing radially outward in a classic stellate habit from a central nucleus. With a hardness of only 2, it is a very soft mineral. It is also very light (specific gravity 2.3–2.4), displays perfect cleavage into plates and shows a pearly luster. Gypsum is used in cement and plaster production.

↑ **Bladed, white barite** with tabular hexagonal prisms of red vanadinite. Not all sulfates and chlorides have high enough solubility to be evaporites. Barium is used to make rubber but it is better known for its application in radiography in the barium meal, a drink containing barium sulfate that fluoresces under X-ray.

EVAPORITE DEPOSITION IN A SALT LAKE

As a mineral lake slowly dries out over a long period of time, the evaporite minerals are precipitated one by one as their concentration builds up in the water. As the lake dries out, the minerals that are deposited leave concentric rings around what were once the edges of the original lake. The lake water must be evaporated to half its original volume before aragonite is deposited. At one-fifth of the lake's volume, gypsum is deposited and at one-tenth volume, carnallite, halite and epsomite.

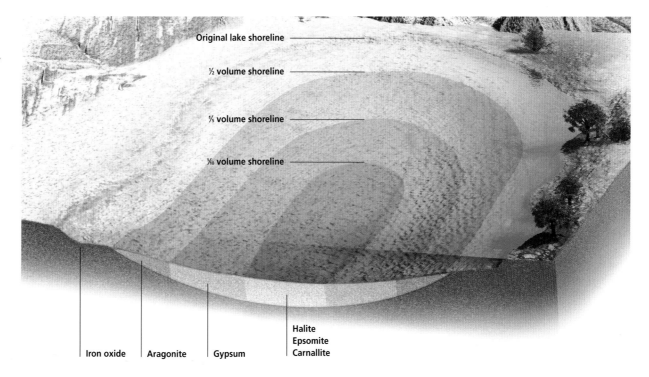

Original lake shoreline ———

½ volume shoreline ———

⅕ volume shoreline ———

¹⁄₁₀ volume shoreline ———

Iron oxide | Aragonite | Gypsum | Halite Epsomite Carnallite

Weathering

Many minerals develop through weathering processes in that thin but important zone where the biosphere interacts with Earth's surface. The water cycle, including seasonal fluctuations of the water table, hydrothermal water and biological activity are all responsible for the minerals in this zone. Key mineral types include clays (montmorillonite, kaolinite) and silica. They are common in different soil horizons depending on the climate zone, and they result from the biochemical breakdown and leaching of other pre-existing minerals and the biological accumulation of new minerals. Of these, some of the most spectacular minerals are the carbonates and sulfates that have grown in the oxidized zone (above the water table) on top of deep ore bodies. Being closer to the surface and easier to find, mine and smelt, these minerals were the first-used metal ores. Some important weathering-zone carbonates are malachite, azurite, cerussite, rhodochrosite and smithsonite. The sulfates include barite, anglesite, chalcanthite and celestine. Both groups yield many important gems.

Scanning electron micrograph of kaolinite shows the tiny pseudohexagonal lamellar plates making up this hydrous aluminum silicate or clay. Weak bonding between the plates explains why, when wet, clay becomes plastic and easy to mold. Kaolinite is used in paper, cosmetics, rubber, china, toothpaste and coffee whitener.

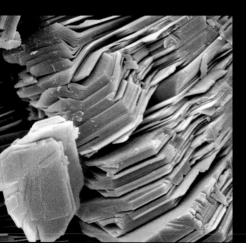

BAUXITIC WEATHERING PROFILE
In the tropics, the seasonal movements in the groundwater level, combined with bacterial activity, lead to weathering profiles such as those found in the far north of Australia. Bauxite (aluminum oxide) and laterite (iron oxide) accumulate in the zone between the high and low water levels, forming red–brown spherical nodules known as pisolites. This buildup of iron and aluminum in the soil gives these tropical bauxite- and iron-rich landscapes their vibrant rusty orange color.

The brilliant orange cliffs of Gunner's Quoin, Gurig National Park, Northern Territory, Australia, are composed of weathered bauxite, the ore of aluminum.

↓ **Smithsonite displaying** a teardrop habit. When massive with good color and banding, this zinc carbonate is cut as a gemstone. Smithsonite is a lesser ore of zinc.

⤋ **Azurite aggregate** is made of tiny copper carbonate crystals. It is used as an ornamental stone.

⤋ **Malachite from Burra,** South Australia, showing light and dark concentric growth bands. This copper carbonate is an attractive ornamental stone and an ore of copper.

This rare rhodochrosite transparent pink hexagonal crystal is a semiprecious gemstone. A manganese carbonate, rhodochrosite is a lesser ore of manganese.

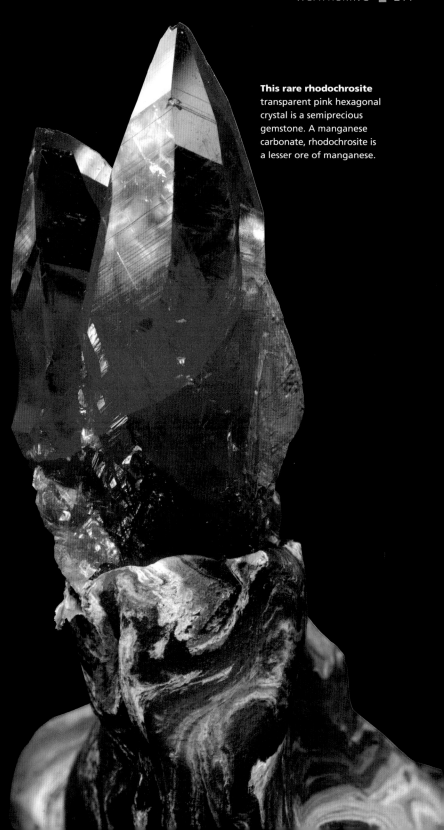

Silicates

The silicates make up the most widespread group of minerals, and account for about 90 percent of all minerals found at Earth's surface. They are the building blocks of rocks. Silicates are classified according to their structure rather than their chemistry. Their own basic building block is a tetrahedron, made up of four negatively charged oxygen atoms surrounding a central positively charged silicon atom. There is an excess of negative charge from this arrangement, so these tetrahedra combine with positive metal atoms, as well as with each other, to build various structures. These arrangements dictate the general properties displayed by the minerals in each group.

↑ **This olivine species forsterite** is from Kagan Valley, Pakistan. Olivine is a hard (6.5–7) neosilicate. Large, transparent, well-formed, orthorhombic crystals such as these are often faceted as gemstones. Forsterite is the most abundant mineral in Earth's mantle.

↗ **Chiastolite is a variety of andalusite** containing fine, black, carbonaceous inclusions arranged in a regular pattern. When the ends of the orthorhombic crystals are polished, as in this specimen from Alconie Hill, South Australia, a cross is displayed. Such crosses are used as religious gemstones.

TYPES OF SILICATES

The simplest silicates are the nesosilicates, in which each tetrahedron is an island linked by combinations of metal atoms. These tend to be hard, with high specific gravity and high refractive index, and display blocky or equant habit. The other groups involve the sharing of oxygen atoms between the corners of the tetrahedra. In the sorosilicates, two tetrahedra share one oxygen atom and the basic building block has two silicon atoms and seven oxygen atoms. The cyclosilicates are rings of tetrahedra; the inosilicates are chains; the phyllosilicates are sheets; and the tectosilicates are three-dimensional arrangements of tetrahedra sharing almost all of their oxygen atoms.

Lazurite from Afghanistan displays rare dodecahedral crystal form. A hard (5.5) mineral, it usually occurs as compact, massive aggregates with pyrite and calcite. Best known as lapis lazuli, this prized gemstone was widely used in ancient Egypt.

MINERALS OF EARTH'S CRUST

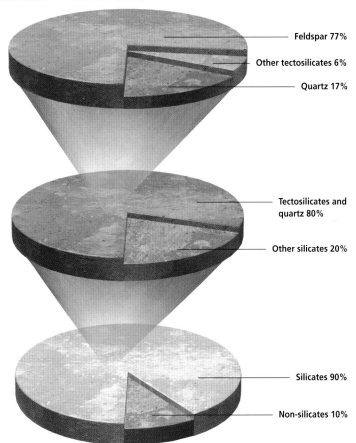

Feldspar 77%

Other tectosilicates 6%

Quartz 17%

Tectosilicates and quartz 80%

Other silicates 20%

Silicates 90%

Non-silicates 10%

↑ **Silicates are the most** abundant minerals. These rock-forming minerals make up the vast majority of all surface rocks. Of these, feldspar is the most abundant mineral overall. Other common silicates are muscovite, biotite, hornblende, augite and olivine. Quartz—silicon oxide—is also abundant.

Blue, tabular, trigonal prisms of benitoite with black columnar neptunite occur here in a massive aggregate of natrolite. Blue benitoite is often cut as a gemstone but it is particularly rare, being found only in San Benito County, California, USA.

Heulandite displays tabular habit. This is a particularly light, fragile, moderately hard (3.5–4) zeolite mineral. It displays perfect cleavage and will dissolve in hydrochloric acid. It forms in cavities in volcanic rocks, especially in basalt.

Feldspars

The feldspars are tectosilicates and comprise the most abundant group of minerals in Earth's crust—they occur in almost every rock type. Although feldspars are mechanically strong, they tend to break down chemically, forming clay, particularly in tropical climates. Where feldspars occur, they leach nutrients such as potassium, which fertilize plants, into the soil. Many feldspars reflect light: sunstone sparkles; moonstone has a cloudlike sheen; and labradorite changes from one vibrant peacock hue to another. Other feldspars are microcline (including the green variety, amazonite) and orthoclase (usually white). Feldspars are used mainly in ceramics, but some varieties are cut and polished as gemstones.

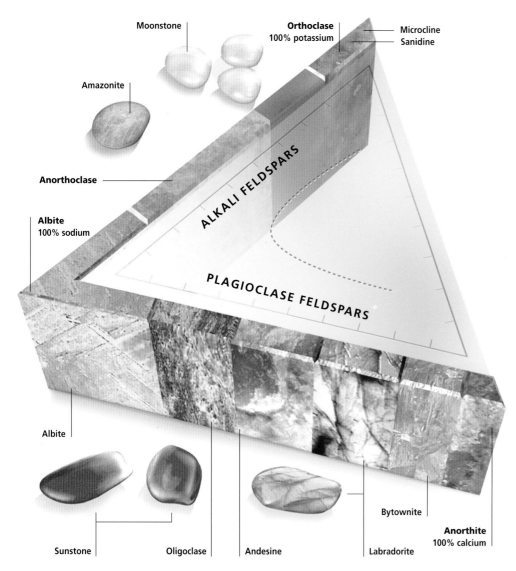

Moonstone

Orthoclase
100% potassium

Microcline
Sanidine

Amazonite

ALKALI FELDSPARS

PLAGIOCLASE FELDSPARS

Anorthoclase

Albite
100% sodium

Albite

Sunstone

Oligoclase

Andesine

Bytownite

Labradorite

Anorthite
100% calcium

FELDSPAR COMPOSITIONAL TRIANGLE

All feldspars are composed of varying amounts of potassium, sodium and calcium. The feldspar compositional triangle is used to indicate the different amounts of each of these elements. A feldspar's position in this triangle determines what type of feldspar it is. The triangle's corners, or end members, are pure potassium, sodium and calcium. Because sodium and calcium atoms are nearly the same size, they substitute readily for one another, so there is a series between these end members. This is known as the plagioclase series and its members include labradorite, oligoclase, albite and anorthite. Sodium and potassium differ in the size of their atoms, so a series exists only at high temperature. The potassium feldspars include sanidine, adularia, orthoclase and microcline. Potassium feldspars often contain numerous veinlike layers of albite, forming what is known as perthitic texture. No series exists between the potassium and calcium end members because their charge and atomic size are so different that they cannot interchange under any conditions.

← **Moonstone comprises** two different feldspars: transparent orthoclase, and albite, which forms planes within the crystal. The adularescence, or schiller effect, it displays is due to their different refractive indexes.

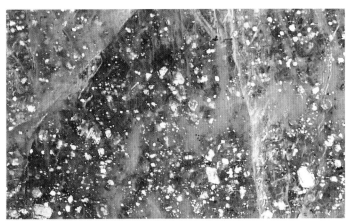

← **The sparkling sheen** of sunstone is very similar to that of aventurine quartz but its background color is a golden honey, orange or red. The sparkle is created by light reflecting off the tiny inclusions of goethite or hematite within the transparent feldspar oligoclase. A variety of sunstone with copper inclusions is found only in Oregon, USA.

↑ **Sodalite is a hard** mineral (5.5–6) that displays a compact massive habit. Cut and polished, it is used as an ornamental gemstone. It is a feldspathoid, a distant cousin to the feldspars. While both are structurally similar, the feldspathoids contain less silica, so they form only in undersaturated rocks— those where there is insufficient silica available to make feldspar; that is, they contain no primary quartz. Other feldspathoids include lazurite, nepheline, leucite and petalite.

→ **These blocky, well-formed crystals** of green amazonite, together with prismatic smoky quartz, are from a pegmatite near Lake George, Colorado, USA. Amazonite is a variety of potassium feldspar known as microcline. It is hard (6–6.5), light (specific gravity 2.5–2.6), with two distinct cleavages at near right-angles. It often displays a perthitic texture (whitish lines).

Crystalline quartz

Quartz is allochromatic: that is, trace impurities will create a variety of colors. Pure quartz or rock crystal is colorless and transparent. The colored varieties are called amethyst (violet–purple), citrine (yellow), smoky or cairngorm (brown), morion (black), rose (pink), sapphire quartz (blue) and milky (semi-translucent white). Milky quartz results from the inclusion of numerous tiny gas bubbles. Aventurine quartz displays a sparkling sheen (aventurescence) from included plates of fuchsite mica or goethite. Tigereye occurs when quartz replaces the amphibole, fiber for fiber. Quartz is a silicon oxide, a three-dimensional network of silicon and oxygen atoms, crystallizing as columnar hexagonal prisms. It is hard (7), has no cleavage but breaks with a conchoidal fracture. Like feldspar, quartz is extremely abundant. Being mechanically and chemically resistant, it will not break down in most environments and, as a result, it is a primary constituent of most clastic sedimentary rocks, as sand- and silt-sized grains. Quartz has a wide variety of important uses: glass and ceramics manufacture, cement and construction applications, and in electronic and optical applications. Despite these practical applications, it is perhaps best known as a semi-precious gem and ornamental stone.

↑ **This colorless and transparent** rock crystal quartz from Diamantina, Brazil, displays the typical habit of six-sided columnar crystals with pyramidal terminations.

↓ **This Brazilian amethyst geode** is filled with crystals growing inward from the wall. Geodes like this can grow extremely large—a European gem dealer once used one as an office.

↖ **Rutilated quartz** forms when thin hairlike fibers of randomly oriented yellow rutile are included in the crystal as it was growing. This gem material is sometimes known as grass stone.

← **This smoky quartz** in well-formed hexagonal crystals is from the subpolar Urals in Russia. These crystals display horizontal striations on the prism faces and have pyramidal terminations formed by positive and negative rhombohedra.

↑ **Two clusters of translucent pink rose quartz** occur on a pegmatitic matrix of platy green muscovite and white feldspar. This specimen is from the Shigar Valley in Pakistan.

→ **Faceted amethyst and citrine.** Two of the most popular colors of quartz are amethyst (purple) and citrine (yellow). A particularly rare variety from Bolivia known as ametrine displays both colors in the one crystal.

Cryptocrystalline quartz

Cryptocrystalline quartz or chalcedony is made up of masses of tiny interlocking microcrystals of quartz. It occurs in an astonishing variety of colors and patterns, many of which have been given a wide variety of unofficial names by local collectors. Chalcedony precipitates from silica-bearing groundwater running through open cracks and channels in the rock. Colors include yellow, blue, orange–red (carnelian), reddish brown (sard), bright green (chrysoprase) and dark green (plasma). Bloodstone or heliotrope is a form of plasma with numerous blood-red spots.

The term agate is used when different colored chalcedonies form layers, generally concentrically, following the contour of a cavity or geode. Blue lace agate comprises delicate alternating bands of translucent mauve–blue and white. Moss agate is translucent chalcedony with fossil-like green dendritic inclusions of chlorite. Ruin agate occurs where agate bands have been shattered by earth movement and then re-cemented by a chalcedony of a different color. Onyx is black-and-white banded agate, while sardonyx displays brown-and-white bands. The infinite textures of agate are best displayed in polished slabs.

↑ **These thunder eggs** formed when cavities in rhyolite lava filled from the bottom up with horizontally banded agate. One (at right) has only partially filled with agate, the rest filling with quartz crystals.

→ **In a petrified forest** in Arizona, USA, buried trees were replaced by cryptocrystalline quartz. Everything is perfectly preserved including growth rings and cell structure. The stone is

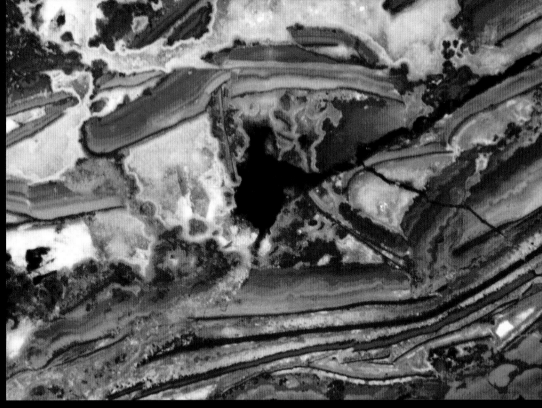

This bowl is made from carnelian, the stunning orange–red variety of cryptocrystalline quartz. It is translucent, hard (6.5–7) and very tough because of its interlocking crystal structure. This gemstone has been used for carving and in jewelry since Biblical times.

This jasper "ruin agate" was formed when bands of opaque red and yellow jasper fractured and were disrupted by some form of earth movement. Later the cracks were filled in with a white translucent cryptocrystalline quartz.

Mexican agate is formed by the precipitation of successive concentric layers of different colored cryptocrystalline quartz from silica-rich fluids filling open cavities in rock.

Chrysoprase is a bright apple green variety of cryptocrystalline quartz colored by nickel. Prized for carving and beads it is sometimes erroneously called Australian jade.

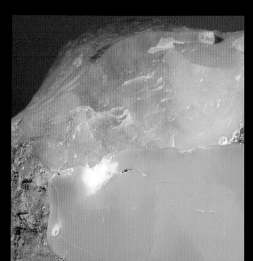

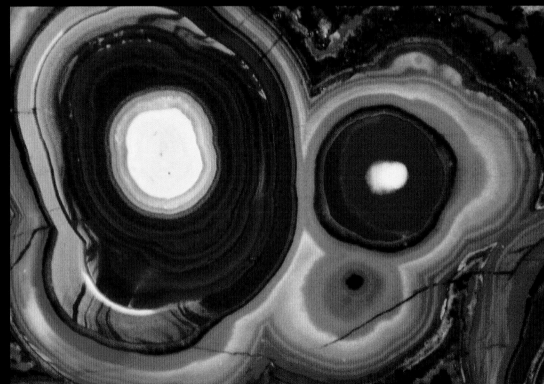

Opals

Opal is quite different from other minerals because, rather than being crystalline, it is an amorphous material that is deposited from silica-rich solutions under certain conditions. Opal precipitates in cavities as nodules or botryoidal masses, or fills cracks as thin sheets. It may even be found replacing bones, shells, wood or other minerals. It is unclear what causes the tiny silica spheres that make up opal to pack together in an orderly manner to form precious opal, or in a disorderly fashion to form common opal or potch because both types are often found side by side on the opal fields. An opal is valued according to its body tone, which can vary from black to light, transparency and play-of-color. A black opal that is free from flaws and inclusions, and displays a uniform distribution of bright spectral colors, can be more valuable per carat than a diamond. On the other hand, a piece of white, opaque, common opal displaying no play-of-color has no value. Good Mexican fire opal displays play-of-color on top of a transparent red body color. Australia supplies most precious opal to the world market.

↓ **The Galaxy opal** is a huge 1.15 pound (523 g) opal discovered in 1989 near Jundah in outback Queensland, Australia. It exhibits the most desirable play-of-color; that is, one that covers the entire spectrum from violet to red.

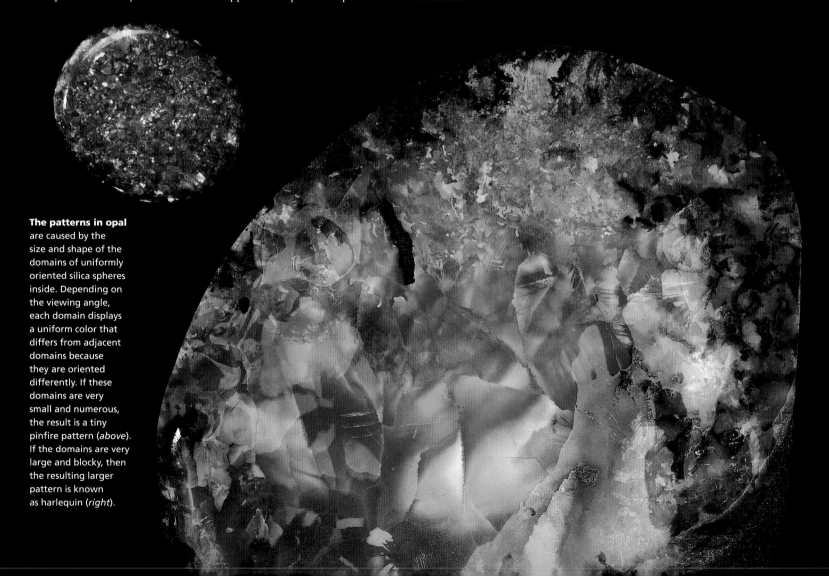

The patterns in opal are caused by the size and shape of the domains of uniformly oriented silica spheres inside. Depending on the viewing angle, each domain displays a uniform color that differs from adjacent domains because they are oriented differently. If these domains are very small and numerous, the result is a tiny pinfire pattern (*above*). If the domains are very large and blocky, then the resulting larger pattern is known as harlequin (*right*).

Precious boulder opal from Queensland, Australia, shows blue-to-red play-of-color, and a body tone that varies from dark (top and bottom right) to light (top and center left). A dark body tone displays the play-of-color to best effect.

Hyalite or glassy opal is a valuable variety of transparent opal that, like precious opal, also shows play-of-color. It is sometimes erroneously called crystal opal. Hyalite usually occurs in cavities as stalactitic, botryoidal or reniform masses.

Moss or dendritic opal is a semi-translucent to translucent variety of opal or potch. It contains plantlike growths of black manganese and iron oxides. This type of opal is polished to display the dendrites and is sometimes set in jewelry.

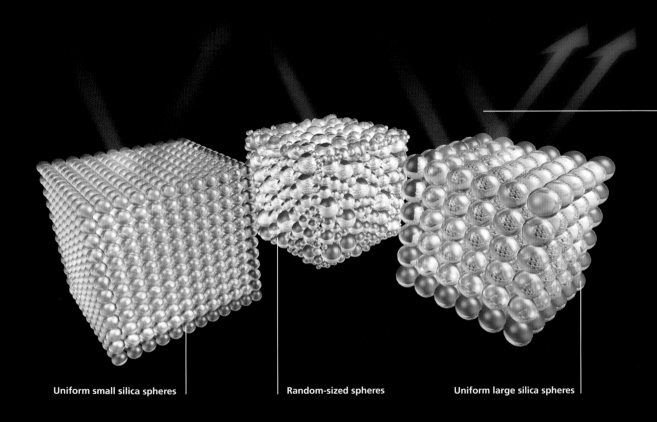

White light

Uniform small silica spheres

Random-sized spheres

Uniform large silica spheres

PLAY-OF-COLOR IN PRECIOUS OPAL Electron microscope studies have revealed that play-of-color in opal is caused by the diffraction of light by a "diffraction grating" of rows of tiny silica spheres. Sphere size determines how much of the spectrum is visible, with the smaller spheres showing only blue play-of-color.

Diamonds

Diamonds are the most romantic of precious gems and the most stable in price, as a result of world marketing and careful stock control. The hardest natural substance, diamonds are high-pressure carbon that has crystallized in Earth's mantle at depths of over 90 miles (145 km). They travel upward in a matter of days in rare magmas called kimberlites or lamproites, which burst out at the surface with a supersonic explosive blast. The resulting carrot-shaped pipes and the surrounding rivers are mined for their diamonds. Rare and most prized are the fancy colored diamonds, especially the intense red. More common but less attractive, industrial-quality stones called carbonado or bort are used in the manufacture of drills, saws, abrasives and polishes. The lineage of some famous diamonds can be traced back for centuries. The Koh-i-noor (Mountain of Light), first reported in 1304, has made its way through many hands and is presently in the British crown jewels.

→ **Cut diamonds** display a range of fancy colors. These command the highest prices, as the majority of diamonds are colorless to pale yellow or brown.

↓ **Major diamond deposits** are generally associated with explosive volcanic pipes in the oldest areas of the continents (called cratons), shaded purple.

THE WORLD'S LARGEST DIAMONDS

Some famous diamonds are here compared to a one-carat engagement-ring. The 530.2-carat Star of Africa (Cullinan I) is the world's largest. The flawless Millennium Star is the sixth-largest gem-quality diamond and took three years to cut with lasers. The Regent diamond was found in in 1701 by an Indian slave. The Blue Hope diamond, now residing in the Smithsonian Institution, USA, is reputed to be notoriously unlucky for its owners.

← **The octahedral face** of this diamond crystal is enhanced with false colors to show the typical triangular-shaped pits. These were etched by hostile chemicals while the diamond was on its way to Earth's surface.

↓ **A scanning electron micrograph** shows how the tip of a dentist's drill is impregnated with fine, angular diamonds. Its hardness of 10 is no match for human teeth, which are made of apatite, which has a hardness of 5.

Star of Africa 530.2 carat

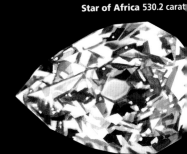

Millennium Star 203 carat

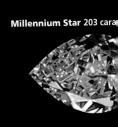

Regent 140.5 carat

Blue Hope 45.5 car

Ring 1 ca

Emeralds and other beryls

Emerald is the best-known and most highly prized intense green variety of the mineral beryl. A large gem of good color and quality may be worth more than an equivalent-sized diamond. Colombia supplies about 50 percent of the world's emeralds, and it was here that the largest gem-quality crystal of emerald was found in 1961, weighing 7025 carats (about 3 pounds or 1.4 kg). Colombian emerald has been steeped in legend and danger since the Conquistadors arrived in the late 13th century. They found the indigenous people mining and using this stone for religious offerings, personal adornment and trade. They soon had them enslaved and mining emeralds on a large scale. There are many other colors of beryl, each with its own special name: aquamarine (light blue to greenish blue), golden beryl (intense yellow), heliodor (light yellow), morganite (light pink to pinkish orange), bixbite (red) and bazzite (blue). Completely pure beryl is colorless and is known as goshenite.

BERYL PROPERTIES

Beryl is beryllium aluminum silicate with a hardness of 7.5–8 and a specific gravity of 2.6–2.9. It is transparent to translucent, with a vitreous luster and white streak. Beryl has an indistinct cleavage and breaks with an uneven to conchoidal fracture. It typically occurs as well-formed, hexagonal prisms with terminations. Chromium produces an intense green color but also strains the crystal. As a result, emerald is quite brittle and the emerald cut, which is rectangular with the corners cut off, was developed to limit breakages.

These beryl gemstones have been fashioned into emerald, round and cushion cuts. They range in weight from an 11.38-carat emerald-cut golden heliodor to a 90.25-carat cushion-cut blue aquamarine.

Emerald deposits of the world. Emeralds and other beryls grow in pegmatites, with gems such as topaz, quartz and tourmaline or in mica schists. The finest emeralds occur in metamorphic calcite veins at Colombia's Muzo and Chivor mines.

→ **A true emerald green** is easily recognizable and highly prized. This well-formed 2.5 inch (6.5 cm) crystal with a flat termination is of the best color. It is from Takowaja in the Ural Mountains, Russia.

↓ **Pale blue aquamarine,** flat-topped gem prisms are intergrown with muscovite plates and white feldspar from the Hunza Valley, Pakistan.

↓ **This gem-quality, near flawless** tabular peach pink morganite crystal on a matrix of platy albite is from the San Diego region in California, USA.

Sapphires and rubies

Sapphires and rubies—varieties of corundum—are the most popular colored gemstones used in jewelry today. Colorless when pure, corundum is one of those minerals that can occur in a number of disguises, depending on the trace elements that are nearby when it is growing. Ruby is colored red by trace amounts of chromium; sapphires exist in a multitude of colors tinted by various combinations of iron and titanium. Padparadsha (pinkish orange), orange and strong yellow sapphires were rare until recently when a new heating process was developed in Thailand. Relatively valueless greenish stones are heated for several days to above 2900°F (1600°C) with beryllium (in the form of powdered chrysoberyl) to produce these fine colors. This has created a glut of these stones in the market and, consequently, a steep fall in price.

Thailand is the most important ruby and sapphire trading center. Most of the world's production passes through Bangkok, where their true identity is lost. Top colored reds become "Burmese rubies" and top blues are called "Ceylon sapphires" after the original sources.

Corundum is second in hardness only to diamond, has a high specific gravity, vitreous luster and a white streak. It crystallizes in the hexagonal system, with ruby commonly occurring as tabular to prismatic crystals and sapphire as pyramidal or barrel-shaped crystals. The largest uncut ruby is reputedly a 45,000-carat pomegranate red crystal weighing a staggering 20 pounds (9 kg).

↑ **Of the faceted fancy sapphires,** blue is the most common color. Orange and yellow stones were rare until recently.

↓ **Ruby in calcite** from Hunza Valley, Pakistan. The well-defined, tabular, triangular-shaped ruby crystal is a typical habit of those growing in silica-poor, aluminum-rich metamorphic rocks.

Major sapphire and ruby deposits are predominantly heavy mineral accumulations in rivers that drain corundum-bearing rocks. The locations marked with red are ruby deposits and those in blue are sapphire-rich regions.

Montana, USA
Afghanistan
Kashmir
Burma
China
Nigeria
India
Thailand/Laos/Vietnam
Kenya/Tanzania
Sri Lanka
Brazil
Madagascar
Australia

♥ Rubies
♦ Sapphires

← **The asterism or star** effect in the DeLong Star Ruby is caused by light reflecting off tiny oriented rutile needles included in the stone. These are visible only under magnification. Found in Burma in the 1930s, this 100.3 carat ruby is one of the world's largest. In 1964 it was stolen, and returned only after 10 months of negotiations plus a US$25,000 ransom.

↓ **These ruby crystals** (*below*) were mined from concentrates of heavy minerals deposited by rivers that had slowly eroded ancient basaltic volcanoes. Rubies are more difficult to find than sapphires and are generally smaller. Sapphire crystals (*bottom*) are found in similar places to rubies. Both are too hard to be waterworn by rivers—their rounded look is due to chemical etching in the basaltic magma that carried them to the surface.

Tourmalines and garnets

Color is not a very useful property for identifying either tourmalines or garnets as they occur in every shade and hue. Such variety has meant that these minerals are often used in jewelry in place of more precious gemstones. Tourmaline was used in the Russian crown jewels and the beautiful green garnet, demantoid, was a favorite among Russian czars. Most recognizable are the multicolored tourmaline crystal columns, which are reminiscent of exotic layered cocktails in tall glasses. Tourmalines form in pegmatites and other igneous and metamorphic rocks. Garnets occur in different types of metamorphic and igneous rocks as equant crystals and they are typically ball-shaped.

TOURMALINES
From left to right: **Elbaite tourmaline crystals** from Nigeria show multiple parallel growth and diagnostic vertical striations.

Elbaite crystal, from Namibia, cranberry red overgrown on green interior, with bladed purplish lepidolite mica and white albite. It displays a typical simple, flat-topped termination.

Detail of the junction between the two colors in a watermelon elbaite tourmaline. The abrupt color change here is associated with a plane of weakness; others may show a gradual transition.

Cross-sections through a liddicoatite crystal column. The two polished slices are from different heights in the same column and show the complexity of the internal color zonation.

GARNETS
From left to right: **Spessartine garnet** crystal perched on a matrix of smoky quartz, blocky white albite and corroded microcline. This specimen is from Shengus in the northern districts of Pakistan.

Grossular–andradite garnets in skarn, a contact metamorphic rock. These gems started growing as green grossular garnets; as the iron content increased, they finished growing as andradite garnets. This specimen is from the Bathurst district in Australia.

Spessartine garnets from Lechang, Guangdong, China, are equant orange-red crystals, growing on the face of a well-formed crystal of silvery muscovite together with white feldspar.

TOURMALINES

The tourmalines are cyclosilicates. Because its metal atoms are completely interchangeable, tourmaline has an extremely variable chemical composition. This creates a variety of colors and types. Many are used as gemstones, particularly the colors and watermelon crystals of elbaite. Tourmaline colors range from red through green to deep blue and colorless.

GARNETS

Garnets are neosilicates and can be built from a number of metal atoms of similar size and charge, creating a variety of different types, including aluminum, iron, chrome, vanadium and zirconium garnets. Good-quality transparent crystals are often used as gemstones and emerald green demantoids are the most highly prized.

Other gemstones

Hundreds of minerals are used in ornaments or semi-precious jewelry. Here, a handful of some of the more intriguing gemstones are described. Some date back to the earliest civilizations, while others have come to be known only recently—even during the last 30 years. Some are so rare that they are known only from one locality. Others have had unusual applications or have been impostors of the precious gemstones. Some tell an interesting tale. While many jewelry stores will not know of these minerals, there will always be some wholesalers or collectors who do. As most people tend to be interested only in the more common precious stones, the others are often most reasonably priced. Some stones look so much like a diamond, emerald, sapphire or ruby that only a trained gemologist can tell them apart. One such red spinel, originally thought to be a ruby, is the centerpiece of the British Imperial State Crown. Gemstones must always be chosen carefully—a zircon looks in every way as good as a diamond but, being softer, it will not stand up to a lifetime of wear.

BENITOITE

A relatively new mineral, benitoite was discovered only at the beginning of the 20th century at Diablo Range in San Benito County, California, USA, and is named for this locality. Because of its color, it was originally thought to be sapphire. This pale to intense blue barium titanium silicate can be cut into a beautiful gemstone. One of its most distinctive properties is a strong pleochroism, or varying of color, which is visible to the naked eye. In one direction it appears to be blue, yet in another it is completely colorless. Benitoite occurs, together with natrolite, neptunite, crossite and albite, in a metamorphic rock known as blueschist.

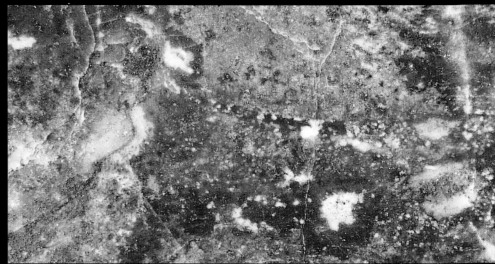

LAZURITE

Lazurite is the key mineral constituent of the metamorphic rock lapis lazuli. Blue lazurite occurs with white calcite and golden specks of pyrite, as seen in this polished specimen from the Flower of the Andes mine in Chile. Lapis lazuli is one of the world's oldest gemstones, referred to in the Sumerian poem *Gilgamesh* and in the Bible. It has been used for jewelry and ornaments since the fifth millennium BC. Most comes from the Sar-e-Sang mine in Afghanistan, which still operates. Historically, lapis lazuli was crushed for a blue pigment called ultramarine. Lazurite is a sodium calcium aluminum silicate with additional sulfate. The sulfate causes a rotten-egg smell to be given off when the material is worked or if hydrochloric acid is applied.

AZURITE AND MALACHITE

Azurite (blue, pictured) and malachite (green) are forms of a hydrated copper carbonate that occur in the oxidized zone above copper deposits. Early sources of copper, these minerals are now prized gems and ornamental objects. Large quantities were used in the Malachite Room of the Winter Palace in St Petersburg, Russia. Azurite will slowly alter to malachite over time. This was noted when blue pigment made from crushed azurite was used in art, only to turn greenish over time, much to the dismay of the artists.

KYANITE

When scratched across the crystal face, kyanite has a hardness of 7, but when scratched along the length of the crystal its hardness is only 4. Kyanite is named after its beautiful color, *kyanos,* Greek for blue. Transparent pieces of good color are cut as gems. However, low-directional hardness and well-developed cleavage limit this use. As one of the aluminum silicate polymorphs, kyanite is an important mineral for determining the temperature and pressure of the formation of metamorphic rocks. It is used mainly in the manufacture of heat-resistant ceramics, such as spark plugs, and acid-resistant laboratory ceramics.

TANZANITE

Tanzanite is the rich, velvety, purplish blue gem variety of zoisite. It is named for Tanzania, the only place where it is found. Tanzanites often display a color change from blue in daylight to violet in incandescent light. Gentle heat treatment brings out the rich blue color. It is believed that Masai herders in the Miralani Hills, southern Africa, were the first to notice this after a lightning strike started a grass fire that heated some ordinary brown zoisite crystals lying on the ground. The deposit was discovered in 1966, and tanzanite was brought to the world's attention in 1969 by Tiffany and Company. Tanzanite and zoisite are members of the epidote group, which also includes thulite, a rose-red gem variety used in ornamental applications.

JADE

The term jade in the gem industry encompasses two quite different silicate minerals of metamorphic origin that have a similar appearance and similar properties: jadeite and nephrite. Precious jade, however, is always jadeite. Because of the extreme toughness of both types, they were used by early civilizations, such as the Olmecs of Mexico, for domestic and ornamental purposes. Jade continues to be extremely popular, particularly in China, for ornamental use, especially for carving the most intricate and delicate objects.

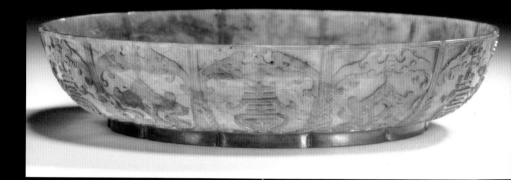

FLUORITE

Fluorite displays the widest range of colors of all minerals—from its well-known purple through to blue, green, yellow, orange, pink, brown and black. With no impurities, it is completely colorless. Some crystals have different colored bands paralleling the crystal surface that reflect several phases of growth under slightly changed conditions. Fluorite is a calcium fluoride, and occurs with lead, zinc and silver sulfides in hydrothermal veins. Although beautiful, it is limited in its use as a gemstone because of its softness and fragility. Enthusiasts consider it a challenge to cut and polish this material. It is used in the steel industry as a flux to aid the smelting of iron ore and also to make fluoride for toothpaste.

OLIVINE (PERIDOT)

Peridot is gem-quality forsterite, a member of the olivine group. Olivine is a mineral series that ranges in composition from forsterite (magnesium silicate) to fayalite (iron silicate). The name olivine comes from the latin *oliva*, in reference to the mineral's olive-green color. It occurs in basalts and gabbros, and is the major constituent of Earth's mantle. Bombs of this material are often blown out during basaltic volcanic eruptions, and olivine also occurs as glassy grains in meteorites. Gem-quality crystals of peridot are not easy to find. Possibly the earliest deposits were mined on the Isle of St John, and peridot was introduced to Europe by the Crusaders. The island, now called Zabargad, is in Egypt.

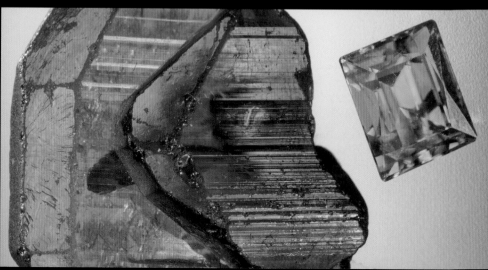

ZIRCON

The ability to split white light into its spectral colors makes zircon a fiery gem. Hyacinth or jacinth refers to the orange–red variety, and jargon is the name given to yellow stones. Jacinth is referred to in the Bible. Zircon self-destructs over time as uranium and thorium substitute for zirconium atoms in the silicate structure. The particles shot out during radioactive decay tear the crystal structure apart, changing it to a non-crystalline metamict state. Such zircon can become cloudy and green, its specific gravity is reduced and it will show no double refraction.

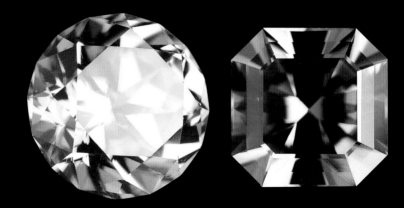

TURQUOISE

Turquoise is a valuable ornamental stone and splendid sky-blue specimens are the most highly prized. Turquoise set in gold found in Egyptian and Sumerian tombs dates back to the fourth millennium BC. At that time it was imitated widely in blue-glazed earthenware; today it is mimicked using powdered copper compounds set in plastics. Turquoise was also used widely by the pre-Hispanic peoples of the Americas, and the turquoise jewelry produced by Navajo silversmiths of New Mexico and Arizona, USA, remains popular. Turquoise (hydrated copper aluminum phosphate) is a secondary mineral, caused by weathering of aluminum-rich rocks containing apatite and copper minerals.

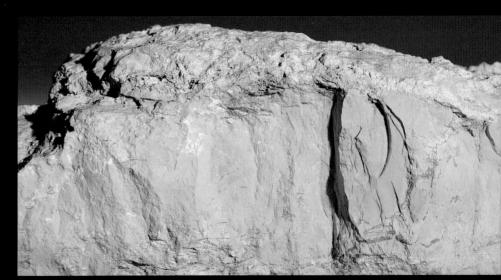

CHRYSOBERYL

Cat's-eye chrysoberyl, or cymophane, is named for its distinctive sheen, a bright line of light reflecting from parallel hairlike fibers within the stone. Colors range from green and brown to the prized golden yellow. Chrysoberyl is a beryllium aluminum oxide that occurs, together with spinel, garnet, tourmaline and beryl, in pegmatites and mica schists. Most valued as a gemstone is the alexandrite variety that shows a remarkable red to green color change. Said to have been discovered on the day the Russian Czar Alexander II came of age, it was named in his honor. Recently chrysoberyl has been used in the heat treatment of sapphire and ruby, as the beryllium it contains causes a dramatic improvement in color.

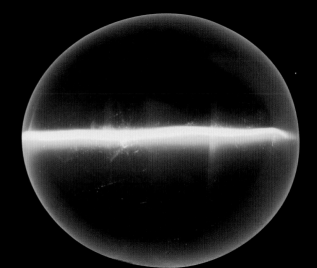

SPINEL

Hard, heavy and durable, gem spinel (magnesium aluminum oxide) is found in heavy mineral deposits concentrated by rivers that drain areas of high-grade regional or contact metamorphic rocks. Often occurring with rubies and sapphires, and being of similar color, spinel was considered one and the same until the advent of modern crystallography. A gemologist can tell them apart by removing the lenses from a pair of polarizing sunglasses and viewing the gem in between—spinel appears dark. A red spinel features in the crown jewels of Britain.

SPODUMENE

Spodumene (lithium aluminum silicate) is a member of the pyroxene family. Its name comes from a Greek word meaning "burned to ashes," referring to its whitish gray color. The gem varieties are kunzite, which displays a pink to violet color due to trace amounts of manganese; and hiddenite, which has a pale to emerald-green color caused by chromium. They were named for American mineralogists George F. Kunz (1856–1932) and William E. Hidden (1853–1918) at the beginning of the 20th century. Gigantic spodumene crystals, sometimes 30 feet (10 m) long and weighing nearly 100 tons (102 t), occur in lithium-bearing pegmatites. Spodumene is mined as an source of lithium.

CALCITE

Calcite, or calcium carbonate, is surpassed in its abundance on Earth's surface only by the tectosilicates and quartz, occurring mainly as marble, limestone or chalk. It displays a wide variety of colors including white, pink, green, yellow, brown, black and bluish. Calcite is used in the manufacture of cement and as a flux. As a coarse crystalline aggregate in the metamorphic rock marble, it is popular for use in sculpture and as a building stone. Calcite has a somewhat limited use as a gemstone because of its softness and well-developed cleavage, so it is more often collected as a mineral specimen. Iceland spar is the perfectly transparent variety that grows in cavities in basalt. It is sought by collectors.

IOLITE

Iolite is the rare gem variety of cordierite. It is a beautiful violet blue. The most distinctive property of iolite is its strong pleochroism. It appears deep blue when viewed down the crystal axis, water clear when viewed from one side, and honey yellow when viewed from the other. Cordierite is a magnesium aluminum silicate that grows in low-pressure regional and contact metamorphic rocks or pegmatites. Iolite crystals were used as polarizing filters by Viking mariners to reduce the effects of haze, mist and cloud while navigating at sea.

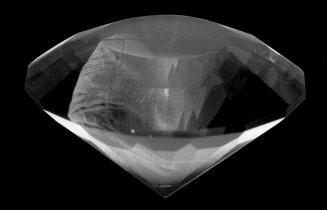

CHAROITE

This rare mineral became known to Europeans only in 1978. It has so far been found in just one locality—along the Chara River in a remote area of east-central Siberia. It is named for the enchanting impression it gives: The Russian word *chary* means "charm" or "magic." Charoite occurs as swirls of spectacular lavender, lilac, violet and purple. Its uniqueness makes it unmistakable, hence appearance is the only test required for accurate identification. It is a hydrated potassium sodium calcium barium strontium silicate and its principal use is for ornamental carvings and as gem cabochons. It forms as a result of metamorphism of limestones by an alkali-rich, syenite intrusion.

TOPAZ

Topaz (hydrous aluminum silicate) is usually highly transparent and can occur in many colors, including the common colorless to pale yellow hues. The rarer colors include intense blue, green, violet, orange, pink and red. Dark orange topaz is known as hyacinth. Some varieties of topaz, particularly the purplish ones, have been heat-treated to enhance their color. A topaz crystal weighing 600 pounds (272 kg), found in Brazil, is the largest known (it is now housed in the Museum of Natural History in New York) and gem-quality crystals weighing up to 220 pounds (100 kg) have been found the Ural Mountains in Russia. Topaz forms in pegmatites together with feldspar, quartz, beryl and tourmaline.

Organic gems

The organic gems are materials that have formed through various biological processes. The earliest examples were bone and teeth, which were employed for utilitarian and decorative purposes by stone-age cultures. Organic materials include teeth, bone, eggs, pearls, amber and kauri gum, vegetable ivory, mother of pearl (pearl oyster shell), paua or abalone shell, tortoiseshell, conches and other shells, red and black coral, ivory and ebony. The latter five of these organic gems have become rare as a result of profiteering and ecologically unsustainable overharvesting, and there is a worldwide ban on any trade in elephant's ivory and tortoiseshell.

DISTINGUISHING FEATURES

Organic gems are light, warm to the touch and much softer than other gemstones. Care must be taken when using them in jewelry, such as rings, or when cleaning them. Pearls are even damaged by body oils, perfumes and cosmetics. The teeth of a number of animals have been used for ivory—the elephant, walrus, hippopotamus, narwhal and sperm whale. In the incessant hunt for ivory, people have even used mammoth and mastodon tusks, known as fossil ivory. Ivory displays internal growth patterns that are distinct to each species. Bone may appear similar but is less dense as it contains a blood-vessel system visible through a hand lens as a series of tiny holes. Tortoiseshell is the horny keratin carapace of the rare hawksbill sea turtle. Jet is a compact, uniform-textured coal formed from the conifer *Araucaria*, while ebony is the dense heartwood of *Diospyros celebica* and displays annual growth banding.

→ **Abalone shell** has an iridescent sheen on its inner surface. The sheen results from light reflecting off numerous oriented microscopic slabs of aragonite which are bound together by a special protein. This beautiful but thin, soft, fragile material is often shaped and inlaid into other materials such as wood and silver. In China, entire sets of furniture are inlaid with patterned iridescent shell.

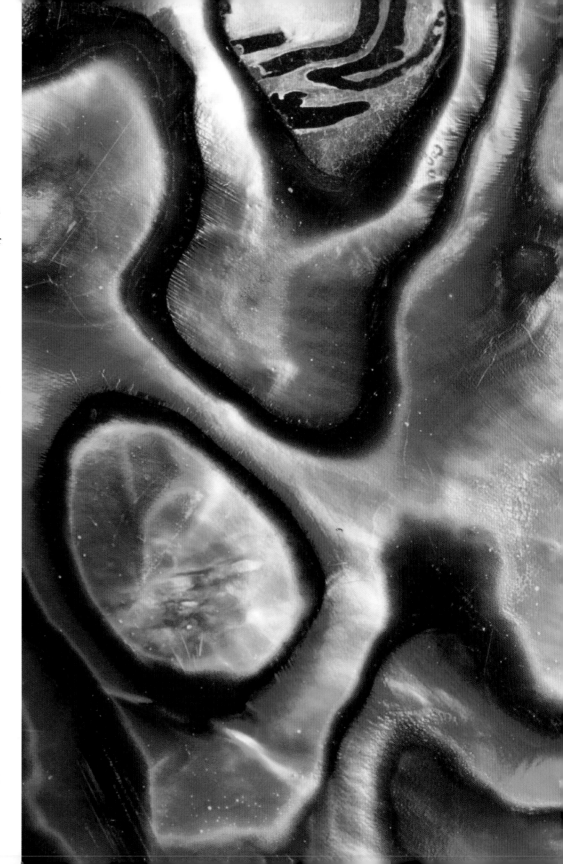

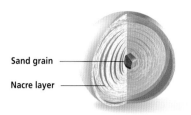
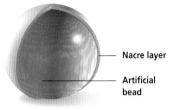

Sand grain —————
Nacre layer —————

————— Nacre layer
————— Artificial bead

Natural pearl   **Cultured pearl**

NATURAL AND CULTURED PEARLS

Natural pearls are made up of layer upon layer of nacre, a substance that is deposited around a grain of sand trapped in the oyster. A natural pearl takes about seven years to form. Cultured pearls are given a head start by inserting a bead within the oyster. This highly successful process was patented by Kokichi Mikimoto in 1896. It takes around three to four years to produce a cultured pearl.

→ **A cultured saltwater** pearl is shown on the inside of the shell of the *Pinctada* mollusk. Pearls are highly prized for their luster and iridescent sheen.

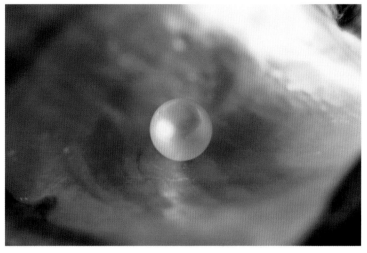

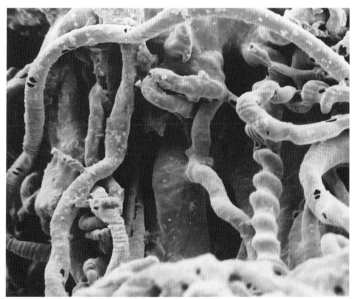

↖ **These pyrite balls** in black shale are from Panawanica, Western Australia. Under certain chemical conditions, iron sulfide will precipitate on organic nucleii and build up to form these golden spheres, which are highly prized by collectors.

← **When gold is** precipitated from groundwater onto the surface of bacterial filaments, sizable gold nuggets can build up over time. These are from Bolivar State, Venezuela. Manganese nodules on the seafloor grow in a similar manner.

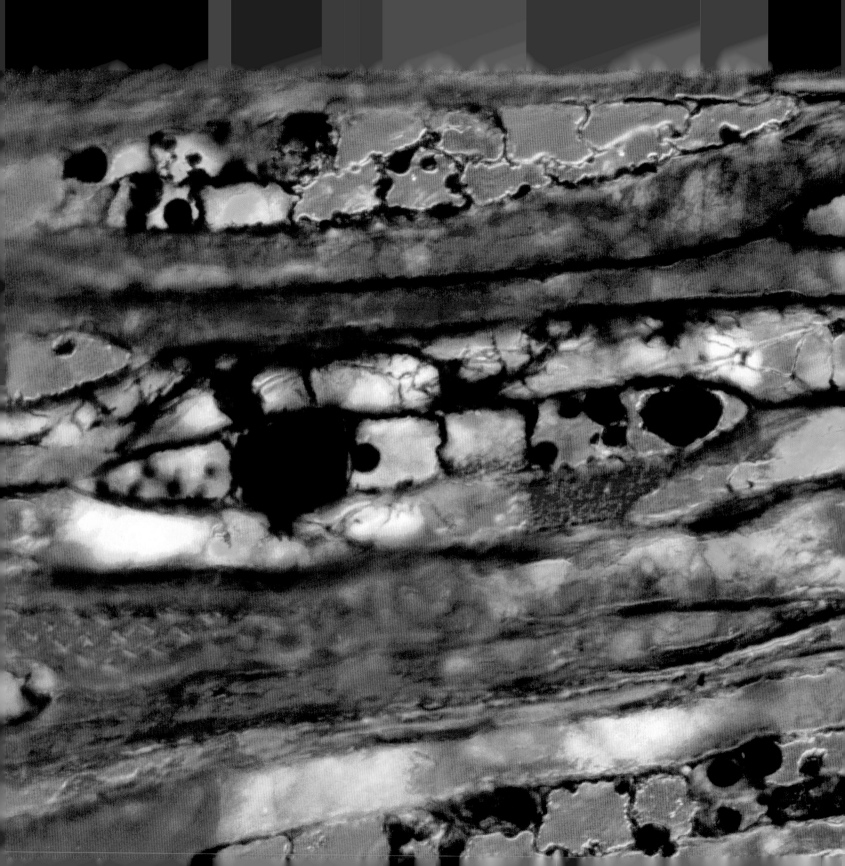

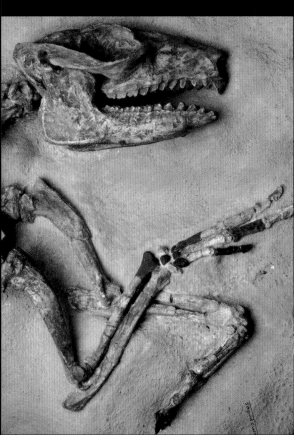

Previous page Wood turns to stone, preserving its cell structure
in the finest detail in this microscopic view of the fossil.

There were signs of abundant life about 540 million years ago. Study and classification of the fossil record, from the single-celled to the most complex organisms, reveal much about life's long history. Fossils also enable geologists to link rocks that may lie continents apart.

Fossil and living groups

Life has evolved like a great branching tree in which all of its different forms are related, having evolved from a common ancestor at the tree's base some 3900 million years back in time. The exact reason or cause of this original event, however, remains unexplained. Throughout most of geologic time, life remained very simple. For over three billion years, life existed in the sea as single-celled organisms, some of which formed into colonies called stromatolites. Only relatively recently, about 600 million years ago, did the tree of life (*far right*) begin to branch into the first soft-bodied multicellular organisms. They began to develop hard parts such as shells, teeth and skeletons about 540 million years ago. Hard parts meant preservation in the fossil record became far more common from this time forward. Plants appeared on land about 410 million years ago. Amphibians, the first land vertebrates, emerged about 340 million years ago. In comparison, the genus *Homo*—humans—appear in the fossil record a mere two million years ago. While some groups evolve quickly, others, such as stromatolites, algae and sponges, remain almost identical to those that lived millions or even billions of years earlier.

THE TREE OF LIFE
The greatly simplified tree of life (*right*) uses lines to show how groups of organisms have evolved through time. Each group can be traced back to see when it arose and to which other groups it is most closely related. Mammals, for example, can be traced in the fossil record back to the late Carboniferous, when they evolved from the stem reptile line. The reptiles can be traced back to the early Carboniferous, when they arose from the amphibians; they, in turn, arose from the bony fishes in the late Devonian. Bony fishes evolved from jawless fishes in Silurian times, and these primitive fishes are found in the fossil record back to the Cambrian. Those lines that fall short of the present, such as trilobites, graptolites and dinosaurs, indicate groups that went extinct. Different line colors indicate phylum-based groupings, most of which appeared during the Cambrian.

↓ **This 50-million-year-old** fossil fish from Wyoming, USA, is remarkably similar to living bony fishes, which developed from primitive jawless fishes during the Silurian.

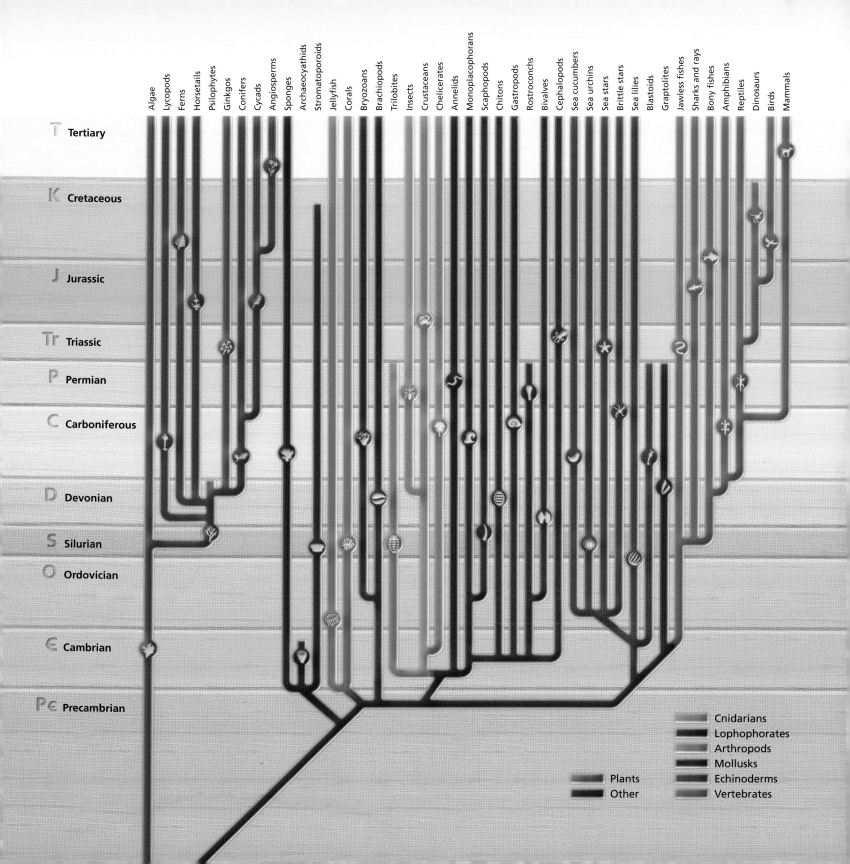

Plants

Plants are classified into groups of increasing complexity, from simple-celled algae to the flowering plants, which have complex reproductive structures. The first land plants appeared in the fossil record during the Ordovician. They were tiny photosynthetic stems carrying simple spore structures on top. Woody tissue, which provides the support needed by larger structures, appeared in the Silurian, followed by ferns in the Devonian. The first flowering plants emerged at the end of the Jurassic. Although plants have soft parts, their thick cell walls make them much better candidates for fossilization than soft-bodied animals. Leaf fossils are common in mudstones and shale, and they are often preserved in delicate detail in these quiet depositional environments. Stronger plant parts such as trunks, cones and seeds can withstand rougher deposition conditions, so these are also found in coarser-grained sedimentary rocks such as sandstone and volcanic ash. In some parts of the world, entire forests have been petrified—literally turned to rock—after being rapidly buried by such sediments.

↑ **Maple leaves** turn yellow and red each autumn and drop to the ground. Only one in a million has a chance to become fossilized. Falling and sinking into the mud at the bottom of a cold, quiet lake will increase its chances.

↗ **This fossilized maple** or sycamore leaf (*Acer trilobatum*) was found at Oeningen, Germany. A deciduous leaf, it has left a carbonaceous imprint in fine siltstone detailing the complexity of its outline and vein structure.

← **These stems and berries** have fossilized, giving a glimpse into the lifecycle of the basswood tree. It is not only the leaves of plants that can fossilize—fruit, flowers, seed pods, pollen and bark are also likely to be preserved. Even tree resin readily fossilizes into the semiprecious gem known as amber.

→ **Petrified forests** occur where mineral-laden groundwater replaces the woody tissue of the buried trees with silica. This process preserves the cellular structure in the finest of detail. Near Sarmiento in Argentina, this entire *Araucaria* forest was growing during the Tertiary when conditions were much hotter and wetter than today. It became preserved when an ashflow buried it. Many of the tree trunks were preserved in their upright position.

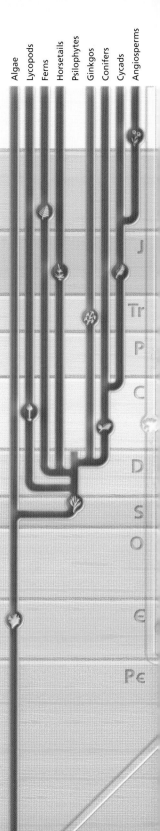

Algae
Lycopods
Ferns
Horsetails
Psilophytes
Ginkgos
Conifers
Cycads
Angiosperms

J

Tr

P

C

D

S

O

Є

Pє

Simple-celled fossils

Simple-celled life falls into two of the five living kingdoms. The monera are the simplest of cells with no nucleus. Instead, they have a loose tangle of DNA strands toward the center, called a nucleoid. Its members, the bacteria, are the most numerous and rapidly evolving living things on Earth. Fossils may form when bacteria sheathe themselves in metals such as iron, manganese or gold as part of the process of transferring nutrients in through the cell wall and sending waste products out. The oldest known fossils are from calcium-carbonate-depositing cyanobacteria. They built the first stromatolite reefs, fossils of which are dated at over 3.5 billion years. These cyanobacteria produced oxygen, which raised atmospheric levels and created the ozone layer. This blocked the Sun's ultraviolet rays, making it safe for other lifeforms to evolve and colonize Earth.

Other simple lifeforms are in the protist kingdom. These are either single-celled organisms, or simple groups of cells, with a nucleus, which are considered neither plant nor animal. Protists that grow hard skeletons or shells may eventually fossilize. These include microfossil groups such as diatoms, coccoliths, forams, dinoflagellates and radiolarians. They can accumulate to form substantial fossil deposits, such as the white chalk cliffs of England. A prolific fossil-forming protist is the reef-building coralline algae. This is made up of fine, threadlike filaments that cover the reef surface and produce calcium carbonate. These trap sand particles and cement them together, contributing significantly to the growth and stability of coral reefs. Eventually an entire reef may fossilize, creating massive hard structures that may be uplifted into vast mountain ranges.

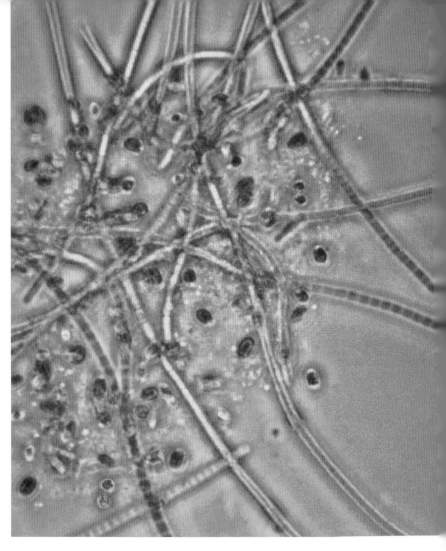

LIVING STROMATOLITE

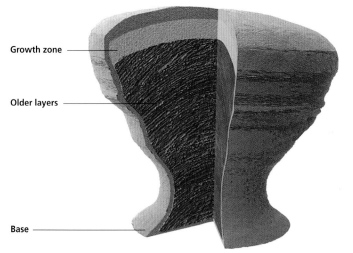

Growth zone

Older layers

Base

FORMATION OF STROMATOLITES

Stromatolites are structures made up of matlike colonies of single-celled cyanobacteria. These microbial mats trap sediment, which becomes cemented by calcium carbonate secretions. The living mat grows upward and outward, forming layer upon layer over time. Its upward growth is limited by reaching tidal level.

↑ **Cyanobacteria are** ancient, aquatic, photosynthetic bacteria that often grew in extensive colonies. Bacteria still make up the largest group of living organisms.

↑ **Living stromatolites** grow upward, layer by layer. At Hamelin Pool in Shark Bay, Western Australia, the stromatolites are over 4000 years old, while fossil stromatolites elsewhere in the state are 3.5 billion years old.

← **A Precambrian fossil** stromatolite from the Pilbara region, Western Australia, clearly shows the curved layers that formed as it grew.

Sponges and their relatives

Sponges, found in the fossil record from the Cambrian to the present, have barely changed during their 600 million years of existence. The most primitive of multicellular animals, they occupy the niche between the single-celled protozoans and the rest of the animal kingdom. Sponges are filter feeders consisting of a simple two-layered body fixed by a holdfast to hard seafloor. They have a wide variety of simple, branching or encrusting shapes. Specialized cells with whiplike flagellum pump water in through pores and out through larger channelways in the sponge. When food particles adhere to their sticky surface, these cells ingest the food and pass the nutrients on to the other cells in the sponge. Sponges may either sexually reproduce as a free-swimming larval stage, or asexually bud new individuals from an adult's body. Whole sponge fossils are relatively rare and it is usually only their tiny, hard, framework-supporting spicules that are preserved in the fossil record. Soft sponges generally do not leave any trace as fossils.

EXTINCT SPONGE RELATIVES

Two important lines of reef-building organisms, which are considered to be types of sponges, both went extinct. The archaeocyathids flourished briefly during the Lower Cambrian and then quickly died out. Their perforated cup-shaped calcareous skeletons were a significant component of the world's first reef structures. The stromatoporoids were also calcareous sponges which formed a hard, compact skeleton. They lived and flourished in shallow reefs during the Ordovician and Silurian periods but were decimated by the Devonian mass extinction. The survivors were finished off by the end-of-Cretaceous mass extinction that also took the dinosaurs.

A giant purple sponge lives alongside a soft coral (seafan) and a yellow crinoid (feather star), off Papua New Guinea's coast. When found alongside other fossils in preserved reefs, sponge fossils enable scientists to piece together entire ancient ecosystems.

These fossil sponges belong to the class Hexactinellida—the glass sponges. All have three-dimensional, six-rayed silica spicules and they cover a wide variety of sponge shapes and sizes. At top center is the vase-shaped *Hydnoceras* species. To its left is a branching variety with a large oscula opening at the end of each branch. On the right is the mushroom-shaped *Coeloptychium agaricoides* from Westphalia, Germany.

Calcareous spicules, which provide support for the sponge *Sycon coronatum*, are seen under a binocular light microscope. Spicules come in many shapes and sizes, and are used to classify sponges.

This fossil stromatoporoid was found on a Devonian reef structure near Golden in the Canadian Rocky Mountains. This class of calcareous sponge was the most diverse before it went extinct.

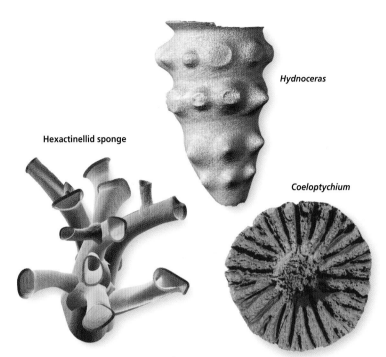

Hydnoceras

Hexactinellid sponge

Coeloptychium

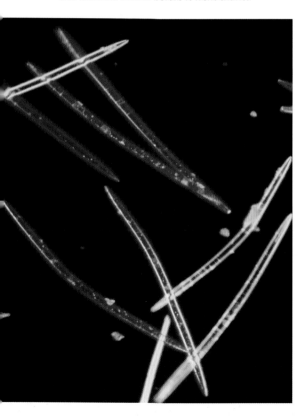

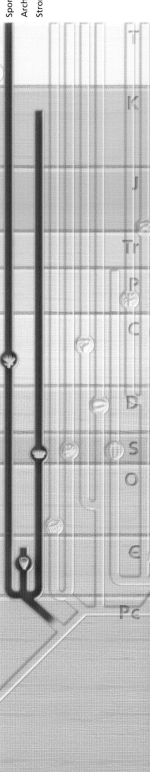

Sponges
Archaeocyathids
Stromatoporoids

T
K
J
Tr
P
C
D
S
O
E
Pc

Corals, jellyfish and anemones

Corals, anemones, hydroids and jellyfish all belong to the phylum Cnidaria, the stinging creatures. Although different in appearance, these simple animals have much in common. One important similarity is their use of feeding tentacles with specialized stinging cells called nematocysts. The two basic cnidarian body forms are very similar—the medusa (jellyfish) and polyp (an upside-down, attached jellyfish). Cnidaria have been around since the Cambrian and hard corals are the most common of all fossils due to easy preservation of their calcium-carbonate supporting framework. Fossils of soft corals, anemones and jellyfish are exceedingly rare. Primitive tabulate and rugose corals flourished along with sponges in the warm seas of the Silurian and Devonian but became extinct 250 million years ago in the global mass extinction at the end of the Permian. Their place was taken by the modern reef-building scleractin corals, which evolved in the mid-Triassic, probably from a soft-bodied anemone-like animal.

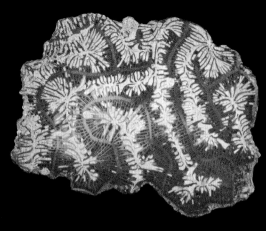

An underwater tropical garden of colonial scleractin (hard or stony) coral slowly deposits calcium carbonate, building a limestone reef that may stretch for hundreds of miles. Fossilized reef structures, preserved as entire mountains of limestone, reveal the evolution and extinction of ancient communities.

↑ **A fossilized Tertiary** brain coral from Antigua clearly displays its long, meandering corallites with numerous, thin, inward-pointing septa.

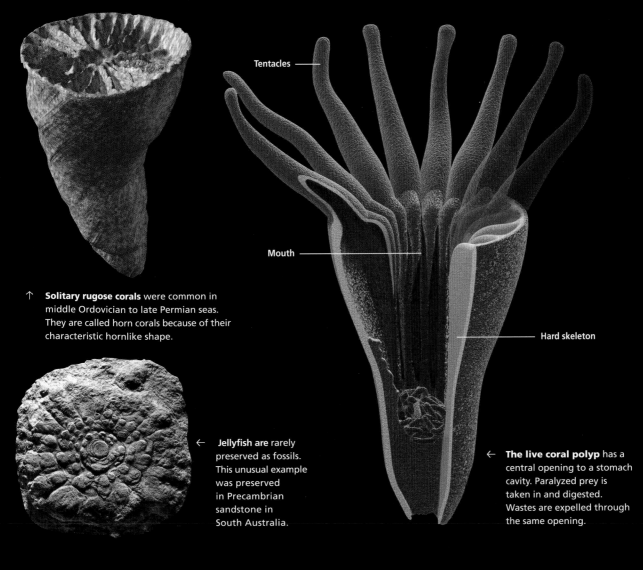

Tentacles

Mouth

Hard skeleton

↑ **Solitary rugose corals** were common in middle Ordovician to late Permian seas. They are called horn corals because of their characteristic hornlike shape.

← **Jellyfish are** rarely preserved as fossils. This unusual example was preserved in Precambrian sandstone in South Australia.

← **The live coral polyp** has a central opening to a stomach cavity. Paralyzed prey is taken in and digested. Wastes are expelled through the same opening.

HOW CORALS BUILD REEFS

Coral polyps remove dissolved calcium carbonate from seawater and precipitate it as the mineral aragonite. This is in the form of a cup-shaped skeleton, called a corallite, in which the polyps live. Thin fins or septa projecting inward from the corallite wall support concertina-like folds in the polyp's gut cavity, giving it a large surface area for digesting food. Virtually all reef-building corals contain zooxanthellae, or dinoflagellates, in the polyp tissue. In return for protection and light, these tiny symbiotic guests provide the polyp with nutrients as the byproduct of photosynthesis. By night, the coral polyp extends its tentacles to feed. Modern hard coral colonies grow in a variety of forms— branching, sheets, plates, encrusting growths and rounded masses known as brain coral. Together with coralline red algae, they create limestone reefs. Extinct fossil forms of rugose coral include a large solitary type (*top left*) known as a horn coral. Its septa are clearly visible in the top of the cup. Some solitary rugose corals reached about 3 feet (1 m) across. They grew by widening the horn upward and building a new living chamber.

Jellyfish

Corals

T

K

J

Tr

C

D

S

O

E

Pє

Brachiopods and bryozoans

Brachiopods and bryozoans are filter feeders known as lophophorates. They have a pair of coiled feeding and respiratory appendages called lophophores covered in fine hairs, or cilia. The beating cilia create a current that draws water through the feeding structure, entrapping unwary micro-organisms. Bryozoans are sometimes called lace corals. Brachiopods are fixed, bottom-dwelling marine organisms enclosed in two shells, which they open for feeding and close for defense with specialized sets of muscles. Their shells are mostly calcareous but some are made of a chitin-based material. Each shell is different and, unlike bivalves, the shells are symmetrical about the centerline of the shell. Brachiopods were usually attached to a solid substrate with a muscular structure called a pedicle, although some fossils had modified shell structures, such as spines or a thick, heavier bottom shell, to keep them upright and stable on a soft sediment bottom. Since their evolution in the Cambrian, brachiopods diversified rapidly to form a variety of shell shapes which have proven to be very useful for dating and correlating rock strata. These include the straight-hinged, wing-shaped spirifers, the spiny-shelled productids, the saw-edge mouthed rhynchonellids, the round, smooth-shelled terebratulids, and the primitive unhinged chitin-shelled lingulids. Most went extinct, the majority being wiped out at the end of the Permian and later mass extinctions. Only relatively few—about two to three percent—of the original genera have survived through to the present. Their evolutionary niche has been filled by the bivalves.

BRYOZOANS

Bryozoans are colonial animals that construct fanlike structures dotted with tiny holes in which individual animals live. The most abundant in the fossil record are those that produce a hard, calcareous supporting framework (*far right*). Bryozoans are more complex than corals, having a three-layered body wall. They also have a separate mouth and anus, and strain food from the water using lophophores. Bryozoans appeared in the Ordovician and they diversified rapidly, making a substantial contribution to building the limestone reefs of that time. Their numbers were nearly obliterated by the mass extinction at the end of the Permian, and their former dominant niche is now occupied by the hard corals.

↓ ***Mucrospirifer arconensis*** was a brachiopod with a very long hinge between the two shells, giving it a winged appearance. Like many other brachiopods, this species disappeared during the mass extinction at the end of the Permian.

← **Archimedes** is the stem of a lacelike bryozoan colony. This spiral-shaped colony is attached to a solid seafloor substrate by the bottom of the screw.

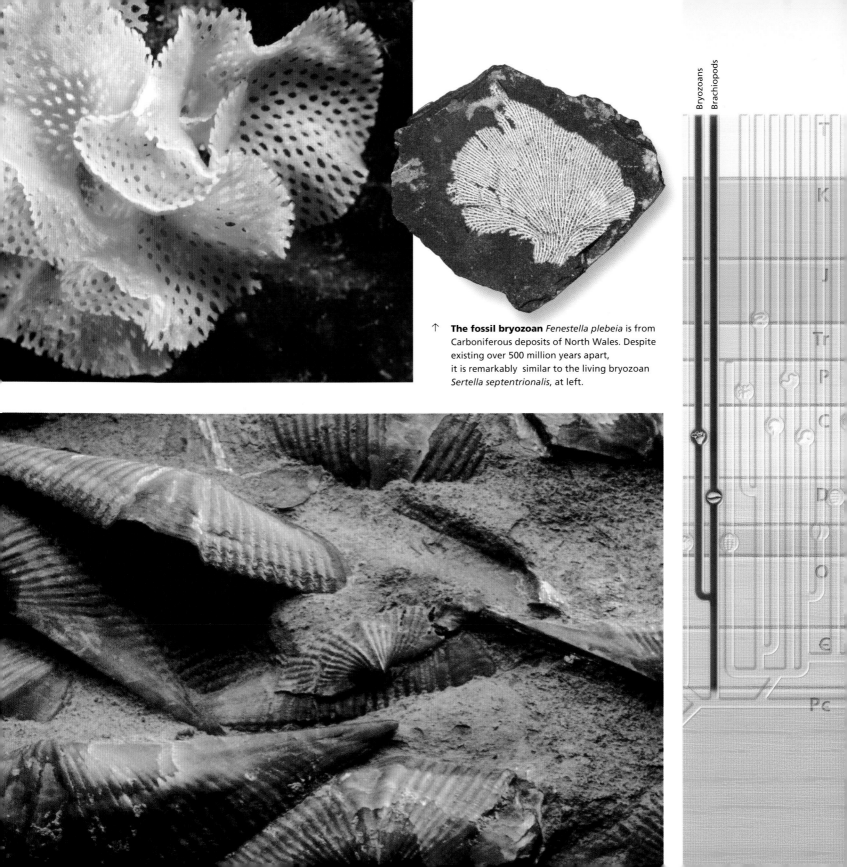

↑ **The fossil bryozoan** *Fenestella plebeia* is from Carboniferous deposits of North Wales. Despite existing over 500 million years apart, it is remarkably similar to the living bryozoan *Sertella septentrionalis*, at left.

Bryozoans

Brachiopods

T

K

J

Tr

P

C

D

S

O

Є

Pc

Mollusks and annelids

Mollusks have proliferated since they first emerged in the Cambrian. Largely marine or aquatic, most secrete a hard calcareous shell that affords protection from predators. Many can close their shells with a watertight seal, enabling them to survive in a variety of conditions. Being hardwearing, their shells often fossilize, and it is not unusual to find masses of fossil shells from ancient mollusks. However, the soft tissue of the animal that lived inside is not usually preserved. Mollusks make up one of the largest phyla in the animal kingdom and include bivalves (clams, mussels), chitons, gastropods (snails, slugs), nudibranchs (sea slugs) and cephalopods (octopus, squid, nautiloids). Annelids are segmented worms, which include marine bristle worms (polychaetes), leeches and the oligochaete worms, such as earthworms. The oligochaetes burrow into soil or other sediment, eating it and digesting any organic material. As annelids are generally soft bodied, they do not readily fossilize, so they are not well represented in the fossil record. They may have evolved from the flatworms during the Cambrian, and probably shared a common ancestor with the arthropods, which include insects, crustaceans and spiders.

↓ **This turret-shaped gastropod** shell has been fossilized in sandstone. Although this mollusk lived as a filter feeder in deep waters where no sand existed, after it died its shell would have washed ashore and later fossilized in the sand.

MOLLUSK FOSSILS

Gastropods and bivalves are two of the most prolific classes of mollusks found in the fossil record. The earliest Cambrian gastropods were straight-shelled and did not coil. Later, they began to coil to form spirals. Gastropods are the only mollusks to have adapted to life on land as snails and slugs. Bivalves developed symmetrical paired shells in the Ordovician. These filter feeders attach themselves to the hard seafloor.

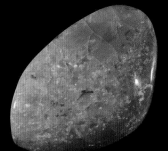

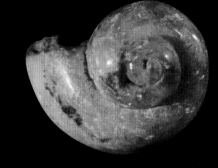

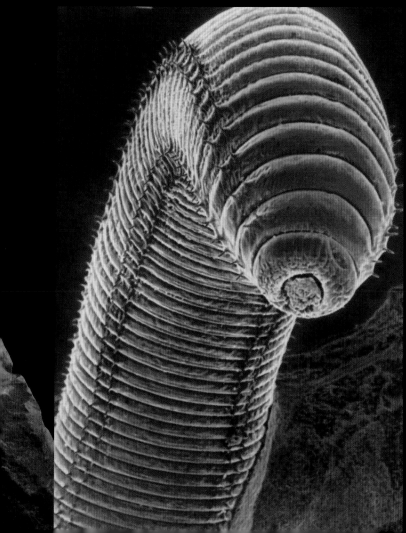

↑ **Opalized bivalves** and gastropods are found in Coober Pedy's opal fields, South Australia. This arid area was covered by a sea in the Cretaceous; now people live underground to stay cool.

↖ **A purple-ringed top shell** uses its muscular foot for locomotion. Shell coiling was a major evolutionary advance for gastropods.

← *Lumbricus terrestris*, the common earthworm, is a soft-bodied annelid burrower. While fossil worms are rare, their burrows often fossilize.

ANNELID FOSSILS
Annelids date to the early Cambrian. Some of the first mineralized tissues ever known were annelid jaw parts and pieces of interlocking body armor. Members that burrow leave traces of their activity in the soft sediment as they disrupt the fine layering. This activity, known as bioturbation, is very common in the marine fossil record. Tubeworms live among coral. Their fossilized remains can be found, often along with mollusks, corals and other reef-builders.

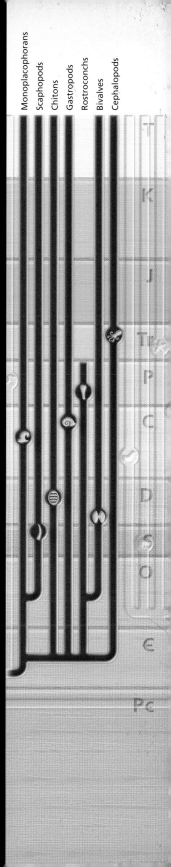

Monoplacophorans
Scaphopods
Chitons
Gastropods
Rostroconchs
Bivalves
Cephalopods

T
K
J
Tr
P
C
D
S
O
Є
PЄ

Cephalopods

Cephalopods are a class of mollusk that includes some of the most important marker fossils. They first appear in the fossil record in the Devonian and have developed and changed rapidly since, surviving several mass extinctions. Cephalopods include three subclasses: the nautiloids (such as the chambered nautilus), the extinct ammonites, and the Coleoidea (octopus, squid and cuttlefish). Ammonites and nautiloids were intelligent marine carnivores that lived much the same way as the nautilus. All the ammonites and nearly all the nautiloids were wiped out 65 million years ago during the end of Cretaceous mass extinction. These animals had complex brains and eyes, used tentacles with sucker pads and had parrotlike beaks to catch and consume prey. They moved rapidly by expelling a jet of water. An animal maintained its buoyancy by pumping gas in and out of its multichambered shells via a siphuncle, a spiral connecting tube.

AMMONITE AND NAUTILOID FOSSILS

Ammonites and nautiloids lived in tropical seas. Abundant fossils are found in fossil limestone reefs. They built straight, coiled or partially coiled shells out of calcium carbonate that vary in size from dwarf varieties a fraction of an inch across to monsters the size of a large truck tire. As the squidlike animal grew, it extended its shell, increased the size of the aperture and periodically inserted a new floor or septum in its living chamber. Their importance as index fossils results from evolution of the dividing wall of these septa and the resultant patterns on the surface of the shell (suture lines). With the passage of time, these suture lines changed from a simple to a more complex zigzag pattern. Hence, whole lineages can be traced through the fossil record. They are a common and highly prized fossil. At Lyme Regis in England, specimens too big to carry lie on the beach, weathered out of nearby Jurassic limestone.

Remarkable similarity can be seen between the 120 million-year-old extinct ammonite *Crioceratites* (*above*) from France and the modern chambered nautilus (*below*).

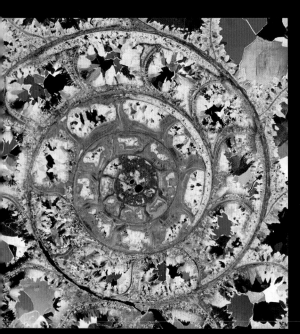

← **Orthoceras,** from Morocco, was a straight-shelled ancestor of the squid. The shell, found in Devonian limestone, has been polished away, exposing its chambers and dividing septa.

← **Growth chambers** in fossil ammonites are similar in structure to the modern nautilus. In this polarized thin section, these chambers have been infilled with calcite and quartz.

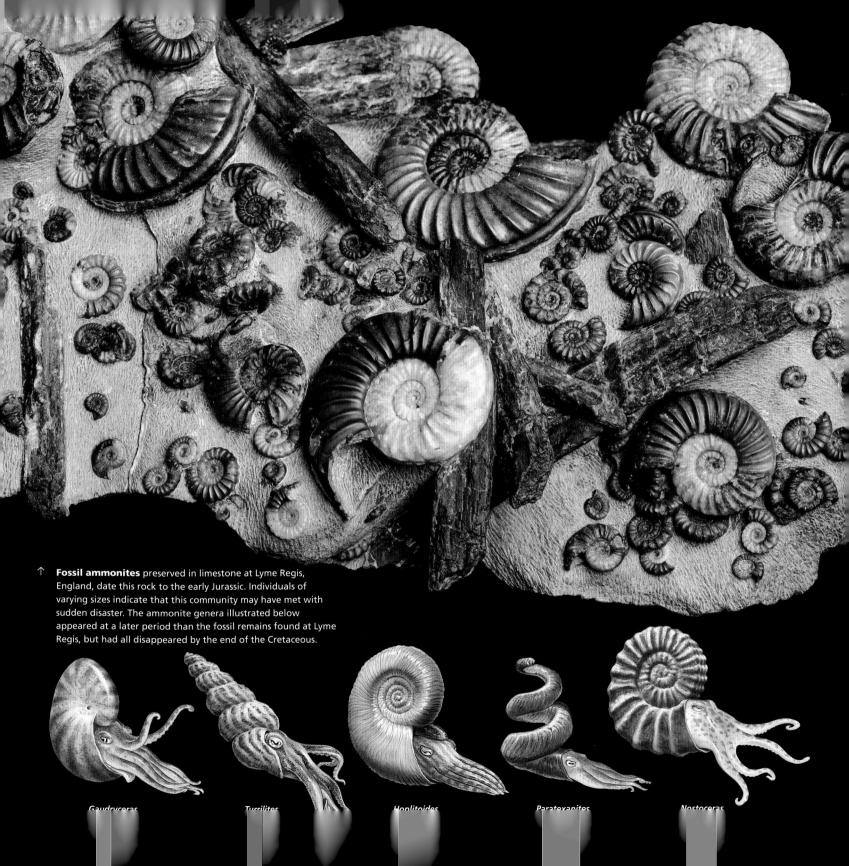

↑ **Fossil ammonites** preserved in limestone at Lyme Regis, England, date this rock to the early Jurassic. Individuals of varying sizes indicate that this community may have met with sudden disaster. The ammonite genera illustrated below appeared at a later period than the fossil remains found at Lyme Regis, but had all disappeared by the end of the Cretaceous.

Gaudryceras   *Turrilites*   *Hoplitoides*   *Paratexanites*   *Nostoceras*

Echinoderms

Echinodermata means spiny skin and is the name given to this phylum as most members have either calcareous spines or interlocking platelets in their skin. For this reason, many members are common in the fossil record. These include sea urchins and sand dollars (echinoids), brittle stars (ophiuroids), sea lilies and feather stars (crinoids) and sea stars (asteroids). The sea cucumbers (holothuroids) are either completely soft bodied or have much-reduced platelets, so they are not often fossilized. Echinoderms all operate by using vascular water pressure in their bodies to inflate and move their hundreds of tiny tube feet. They are either attached filter feeders or mobile herbivores or carnivores. Crinoids and the similar-looking blastoids appeared in the Cambrian then rapidly diversified into the other groups. Although the Permian mass extinction took a heavy toll on every class of echinoderm, all but the blastoids survived to the present. However, their species numbers have never fully recovered.

ECHINOID FOSSILS

Many an angler or diver has felt the painful stab of a sea urchin's protective spines. As these spines normally detach from their ball joint after the animal dies, both may be preserved separately. They were common throughout the Carboniferous and Permian. A central hole in the underside of the shell indicates the position of the mouth. Powerful jaws enabled some species to excavate a cavity in solid rock. Most echinoids are rounded and symmetrical, with a top and bottom but no left or right. They hunt for food by moving over the surface of the seafloor. Other types, the irregular urchins, have a bilateral body symmetry as they evolved a streamlined shape more suitable for burrowing.

Sea stars display the typical five-fold body symmetry that is common to all living echinoderms. The Jurassic fossil sea star *Pentasteria longispina* (*right*) and a modern bright vermillion biscuit star *Pentagonaster duebeni* (*above right*) from Australia, both have five identical arms radiating outward from a center of symmetry.

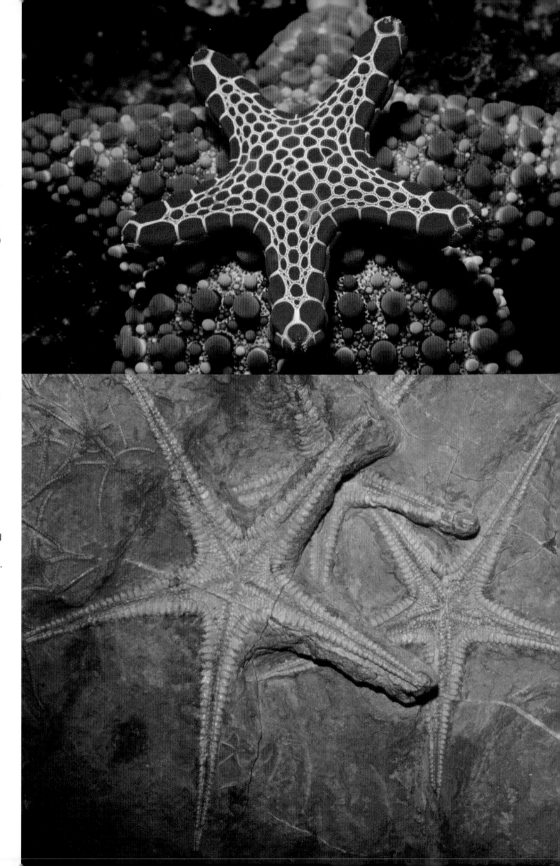

CRINOID FOSSILS

Crinoids were the most abundant echinoderms from the early Ordovician until their decimation at the end of the Permian. Many species had short ranges and thus are useful as local index fossils. The earliest crinoids were fixed forms attached to a hard substrate by a long stem. The later forms had reduced stems and could walk about in search of food. Crinoids were so prolific in places that their remains formed huge thicknesses of limestone. The most common fossils are the disks from their arms, legs and stems, known as columnals.

Like modern crinoids, fossil crinoids have very similar articulated arms joined to a cup-shaped body or calyx. This fossil crinoid preserved in Silurian mudstone was attached to the hard bottom via a long flexible stem of stacked calcareous disks. Modern crinoids lack a stem and move around using articulated legs.

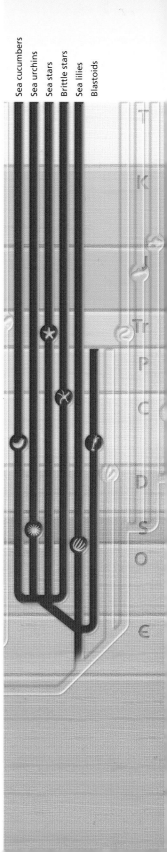

Sea cucumbers
Sea urchins
Sea stars
Brittle stars
Sea lilies
Blastoids

T
K
J
Tr
P
C
D
S
O
€

Arthropods

The arthropods make up the most diverse group of invertebrates. They include the insects (such as silverfish, bees, mosquitoes), crustaceans (crabs, lobsters, barnacles), chelicerates (spiders, ticks, mites, horseshoe crabs, eurypterids) and the trilobites. Arthropods are recognized by their armor-protected bodies and their legs, which are jointed to enable a full range of movement. They are well represented in the fossil record because of this chitinous external skeleton. As an arthropod grows, its shell is regularly discarded and replaced, so an individual's chances of leaving behind a fossil are greatly multiplied. Crustaceans, trilobites and chelicerates all evolved in the Cambrian. Trilobite fossils are attractive and are popular among collectors. They were found and treasured by Native Americans, who called them "water creatures that live in rock." The now-extinct eurypterid, reaching more than 10 feet (3 m) long, and sporting pincered appendages, was a formidable marine predator. The insects evolved from a centipede-like ancestor that lived in the Devonian. Since that time they have diversified to occupy almost every conceivable environment, including the skies.

Agnostida

→ **Eurypterids or sea** scorpions are an extinct group of the chelicerate subphylum of arthropods that lived from Ordovician times through to the Permian. This Silurian specimen clearly shows its segmented body and large flattened pair of legs used for swimming. Eurypterids also had four to five pairs of walking legs and a pair of sharp, pincerlike appendages or chelicerae, used for capturing prey.

WINGED INSECTS
Insects make up the most diverse group, and are the only invertebrates that developed the ability to fly. Despite their abundance, fossils such as this dragonfly (*left*) are comparatively rare because they have such small, delicate skeletons and wings (*below*). Larger insects preserve more easily. Some Carboniferous dragonflies had wingspans of 30 inches (75 cm). The most perfectly preserved winged insects have been trapped and fossilized in tree resin (amber).

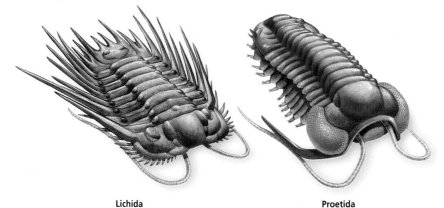

Lichida

Proetida

TRILOBITE FOSSILS

Trilobites were the most numerous and successful arthropods in the oceans during the Cambrian, Ordovician and Silurian. They suffered in some major extinctions, becoming quite rare in the Devonian and Carboniferous. They finally disappeared during the mass extinction at the end of the Permian. Trilobites consist of three sections: the cephalon (head), the thorax (made up of hinged segments) and the pygidium (a tail consisting of fused segments). The basic Cambrian form evolved into other forms. There were, for example, changes in the number of thoracic segments, the development of spines along the shell edge and the evolution of swimming varieties.

Fossils in amber

Hundreds of species of plants and thousands of animals have been preserved in exquisite detail in amber. The animals are usually small arthropods but some larger centipedes and spiders have been found. There have even been rare discoveries of lizards, feathers and mammal hair. To see how a gnat becomes trapped, picture a Tertiary scenario. A gnat alights on a smooth, glassy ledge while searching a coniferous forest for decaying organic matter. Two of its legs become trapped by a sticky resin exuded by the tree to combat a fungal parasite. In the struggle to escape, the fly's legs break off, but a wing and antenna become stuck. Continual struggle draws the fly farther into the resin. It is entombed. Eventually, the tree dies and falls into a muddy lagoon, together with its insect-laden resin. Over time the tree becomes lignite and the polymerization, or molecular bonding, of the hydrocarbon resin causes it to harden. Some 40 million years later, storms erode the coaly layer (known as blue earth) now exposed on the floor of the Baltic Sea. Being lighter than salt water, the amber floats and is washed ashore.

IMPORTANT AMBER LOCALITIES

Most amber comes from the Cretaceous and Tertiary periods. The most significant deposits are from the Dominican Republic (10 to 40 million years old) and the Baltic Sea (30 to 65 million years old). One piece of Baltic amber weighed 20 pounds (9 kg). Older amber is rarer as it breaks down with time, heat or pressure. Some of the oldest amber, found as small fragments in the Carboniferous coalfields of Ayrshire, Scotland, is dated at over 200 million years. It contains microscopic plant and fungal remains. Amber amulets and ornaments have been used since the Stone Age. It was also fashionable among ancient Romans. Some "amber" is faked, using insects set in colored resin.

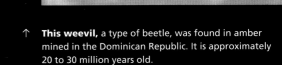

↑ **This weevil,** a type of beetle, was found in amber mined in the Dominican Republic. It is approximately 20 to 30 million years old.

↓ **It is extremely rare** to find a large spider in amber. This is because a larger animal is far more likely to escape the resin's sticky grip.

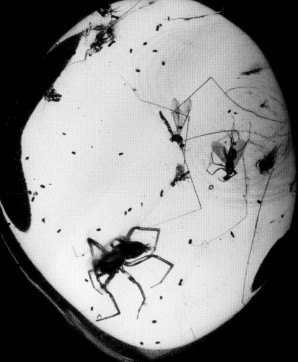

← **Crane flies** became trapped in this Baltic amber about 30 million years ago when they landed on the sticky resin of a conifer tree. They belong to the Tipulidae family of two-winged, mosquito-like insects, all of which have very long legs.

↑ **Fungus gnats** were trapped in the oozing resin of a Baltic conifer tree about 40 million years ago. These gnats belong to the family Sciaridae, and search for fungus or decaying vegetation in which to lay their eggs. They are characterized by long, segmented antennae and delicate bodies with pointed abdomens.

Fishes

The oldest known vertebrates appeared over 500 million years ago. Small, fishlike animals from the Cambrian Burgess Shale, Canada, and Chengjiang, China, lacked a bony skeleton. By the Silurian, many types of armored jawless fishes (agnathans) occupied all water habitats, fresh and salt. Armored agnathans died out in the Devonian but their boneless relatives, lampreys and hagfishes, survive today. Fishes with jaws (gnathostomes) appeared in the Silurian and quickly replaced agnathans. All the major jawed fish groups evolved by the early Devonian, including spiny-finned acanthodians and armored placoderms (both extinct), chondrichthyans (sharks, skates and rays) and bony fishes (ray-finned actinopterygians and lobe-finned sarcopterygians). Actinopterygians dominate today's seas, lakes and rivers. Sarcopterygians had lungs as well as gills, and a strong bony skeleton; they are believed to have given rise to all later vertebrates—amphibians, reptiles, birds and mammals.

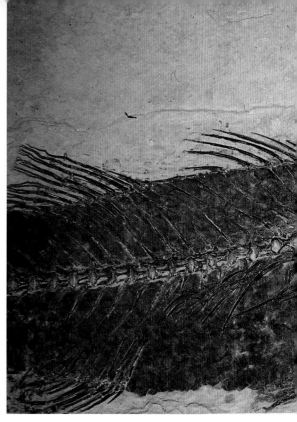

THE COELACANTH: A LIVING FOSSIL

The lobe-finned coelacanth's ancestor is thought to have diversified into all land-dwelling vertebrates. It was known only from fossils until the discovery of a live one in 1938 by Marjorie Courtenay-Latimer. The curator of a tiny museum in a diminutive port north of Cape Town, South Africa, she habitually visited the docks seeking interesting specimens. One day she came across a large, pale mauve-blue fish with iridescent silver markings and paired, fleshy, limblike fins. After the discovery was confirmed by Professor J.L.B. Smith, the mounted specimen, the curator and the professor became celebrities. Submersibles have since found coelacanths living in caves off the east African coast, and there have been recent sightings in Indonesia.

↑ **Disposable teeth** are sharks' most common fossil remains. Their cartilage-based skeletal material is not well preserved.

← **The coelacanth** is a classic "living fossil." Its preferred habitats include submarine caverns at depths up to 2300 feet (700 m). While its numbers are difficult to estimate, it is classified as endangered.

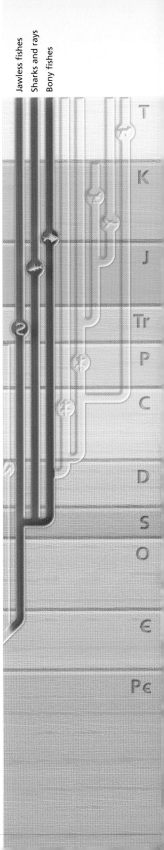

← **The bony fishes** are the dominant aquatic vertebrate today and they make up half of all known vertebrate species. Here, *Mioplosus labracoides*, a voracious Eocene predator, was "frozen" in the act of swallowing a smaller bony fish some 50 million years ago. The pair were buried in limy, muddy lake deposits, which now make up the limestones and mudstones of the Green River Formation in Wyoming, USA. Bony fish remains fossilize more readily than the cartilaginous skeletons of sharks and rays.

← **Fossil bone** from the Devonian bony-headed *Asterolepis ornata*, viewed as a thin section under polarized light, is very similar to that of modern fishes. The typical Haversian canals—the tunnels in the bone for blood vessels and nerves— appear black.

Jawless fishes

Sharks and rays

Bony fishes

T

K

J

Tr

P

C

D

S

O

Є

PЄ

Amphibians

Some ancient amphibians reached an enormous size. Amphibians were the first vertebrate group to venture from the water onto dry land. Amazingly, this evolutionary process, which began in the late Devonian, is mirrored in the lifecycle of of every amphibian. They spend the first part of their lives in the water, swimming and using gills like fishes, then undergo a tremendous body metamorphosis, adapting to life on the land, including the development of lungs to breathe air, like every other land dweller. The best-known example of this process is a tadpole changing into a frog. The tail for swimming gradually shrinks, the body shape changes and legs begin to sprout, which progressively become stronger and more useful. Salamanders undergo a similar metamorphosis, although their body shape does not alter as dramatically and they retain their tails. The timing of this process is controlled by a thyroid hormone called thyroxine. Amphibians will never stray too far from the water as they still need it to reproduce.

ANCIENT FISHLIKE AMPHIBIANS

Some excellent rare examples link the primitive lobe-finned bony fishes with true amphibians. *Eusthenopteron* (*below*) had amphibian-like features. These include rounded, fleshy fins with bones that swivel in shoulder and hip sockets, like the limbs of land animals. It had a similar skull structure, enameled teeth and a connection between the inside of the mouth and the nostrils—common to all land vertebrates. It also breathed air like a lungfish. *Icthyostega* was a fishlike amphibian with well-developed limbs and toes like other land vertebrates, but it had a tail with a large, rayed fin similar to that of a lobe-finned fish. It also had a lateral line organ, as do fishes, to sense vibrations in water. *Icthyostega* probably spent most of its time in the water, using its limbs to move around on the bottom.

Eusthenopteron is a 360 million-year-old lobe-finned fish, considered to be a linking form between bony fishes and amphibians. It had paired, limblike fins and probably breathed with a primitive lung. The lobe-finned line is now only represented by the lungfishes and the coelacanth.

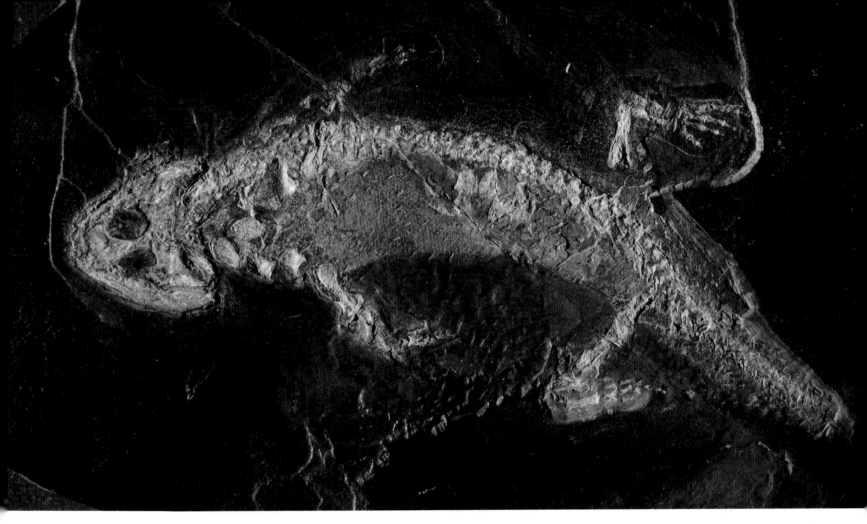

↖ ***Messelobatrachus*** species, a 49 million-year-old fossil frog, was found in Grube Messel, Germany. It is so well preserved that the soft body shape is still visible as a brown outline.

← ***Dendrobates auratus***, the poison arrow frog, has brightly patterned skin to warn predators. Fossils do not reveal the details of such defense mechanisms.

↑ **Salamanders,** such as this *Sclerocephalus* species from Oderheim, Germany, are rare in the fossil record.

← **Like all salamanders,** the yellow spotted newt is an amphibian, not a reptile. All salamanders have thin, moist skin and spend part of their lifecycle in water. A Chinese giant salamander can grow as big as a human.

Reptiles

Reptiles were the first vertebrates to lay amniotic or shelled eggs. These did not dry out like the gelatinous amphibian eggs, enabling reptiles to become less dependent on water and to develop a truly terrestrial lifestyle. They also developed a thicker, scaly, impermeable skin, stronger limbs for better mobility, a more efficient heart and advanced ears. Reptiles appeared during the early Carboniferous and evolved from a group of amphibians called anthracosaurs, of which *Seymouria* is a member. It had an amphibianlike skull while the rest of the skeleton and scaly skin resembled that of a reptile. Reptiles ruled the terrestrial scene for nearly 250 million years, diversifying into many forms, including the dinosaurs.

MARINE REPTILES

The marine reptiles, or members of the subclass Parapsida, are particularly interesting because they were land reptiles that returned and readapted to life in the water. These efficient predators dominated the seas from the Triassic to the Cretaceous. During their 150 million-year-reign, they developed flippered feet and their body shape became more streamlined, fishlike and efficient for swimming and hunting. They included the sauropterygians (long-necked plesiosaurs and strong, short-necked pliosaurs) and the ichthyosaurs.

→ **Aquatic reptiles,** such as this Cretaceous specimen from China, developed streamlined shapes and webbed feet to become efficient hunters.

↓ **Ichthyosaurs** developed from a lizard-shaped to a dolphin-shaped marine reptile. The bones of their limbs fused to form effective flippers.

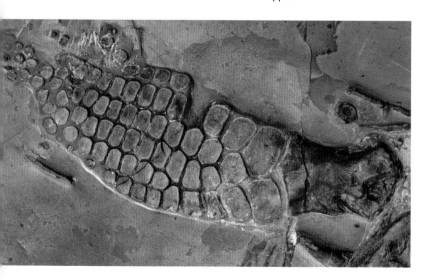

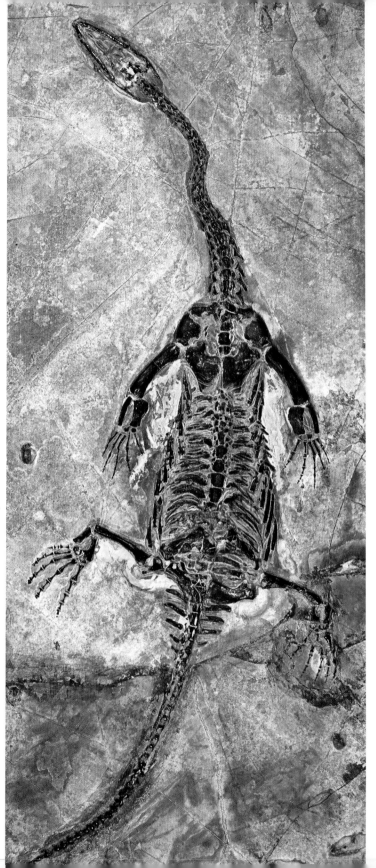

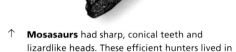

↑ **Mosasaurs** had sharp, conical teeth and lizardlike heads. These efficient hunters lived in the Cretaceous seas 90 to 65 million years ago.

← **Palaeopython,** from the Grube Messel quarry in Germany, is remarkably similar to its living counterpart, adapted for limbless locomotion.

← **Like all snakes,** the jungle carpet python from tropical Australia has a jaw that it can dislocate to swallow prey whole.

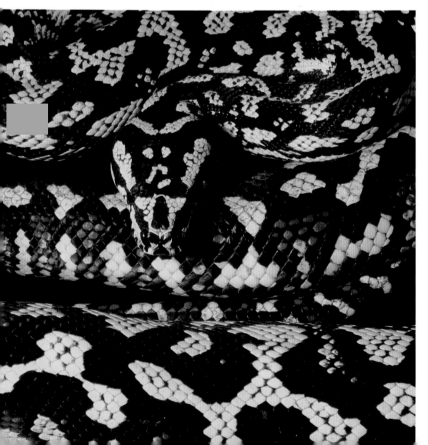

CLASSIFICATION OF REPTILES

As reptiles evolved into effective predators, they needed stronger jaws. Openings known as temporal fossae developed in the side of the skull to allow better leverage for the jaw muscles. Reptiles are classified according to the position and number of these openings. The Anapsida are the most primitive as they have no openings. They include the turtles, tortoises and terrapins. The Parapsida include the aquatic ichthyosaurs and plesiosaurs. The Synapsida are a subgroup that include the pelycosaurs and the strong-jawed doglike beasts known as therapsids. These transitional forms between reptiles and mammals went extinct at the end of the Permian. Diapsida is perhaps the most successful reptilian subclass. It has given rise to the Lepidosauria, which includes the mosasaur and most modern reptiles, and the Archosauria, which includes the dinosaurs, pterosaurs and crocodilians.

Reptiles

Dinosaurs

T

K

J

Tr

P

C

D

S

O

Є

P€

Dinosaurs

Dinosaurs appeared in the Triassic, about 230 million years ago, and ruled Earth for more than 150 million years. When viewed in context with human existence of a mere few million years, their success is apparent. Possibly the main reason for their dominance was the evolution of a hip joint that set dinosaurs apart from other reptiles by allowing them to stand upright, move much faster and grow bigger than anything before then. A right-angled joint at the top of the leg bone allowed their legs to be brought underneath the body, with the body weight balanced over the hips. When the supercontinent Pangea split during the Jurassic, two populations of dinosaurs—one group in Laurasia and the other in Gondwana—began to evolve along separate lines. Diversification continued during the Cretaceous until 65 million years ago, when the dinosaur line was wiped out across the planet.

MEAT-EATING DINOSAURS

Carnivorous dinosaurs evolved many specialized strategies that enabled them to become effective hunters and scavengers. Most importantly, their larger brains gave them the intelligence to develop attack strategies and work in groups to tackle prey bigger than themselves. They had strong jaws with replaceable, sharp, serrated cutting teeth, allowing them to tear and swallow large hunks of flesh whole (*Tyrannosaurus* and *Allosaurus*, *right*). *Deinonychus* had slashing claws on its hind legs. *Baryonyx* used a huge, hooklike claw on its hand to spear fish. *Tyrannosaurus*, the largest of all the predators, weighed more than an African elephant. By far the majority of dinosaurs, however, were plant eaters.

↓ **Twin-crested *Dilophosaurus*** was a fast-moving, slender, meat eater that walked upright on two muscular legs. It hunted in packs 201 to 189 million years ago, during the Jurassic. The twin, semicircular crests may have been a mating display.

← **Dromaeosaurus** fossils from Liaoning province in China, such as this juvenile, were covered with featherlike fibers and filaments. Such discoveries provide important details about the evolution of birds from the dinosaur line.

Tyrannosaurus
39 feet (12 m) long

Saltasaurus
39 feet (12 m) long

Corythosaurus
33 feet (10 m) long

Euoplocephalus
23 feet (7 m) long

Pachycephalosaurus
26 feet (8 m) long

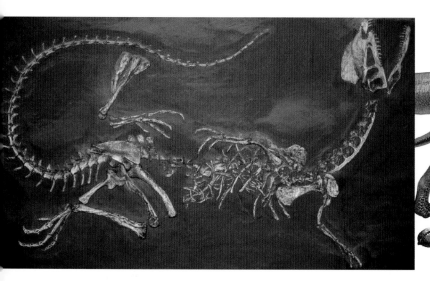

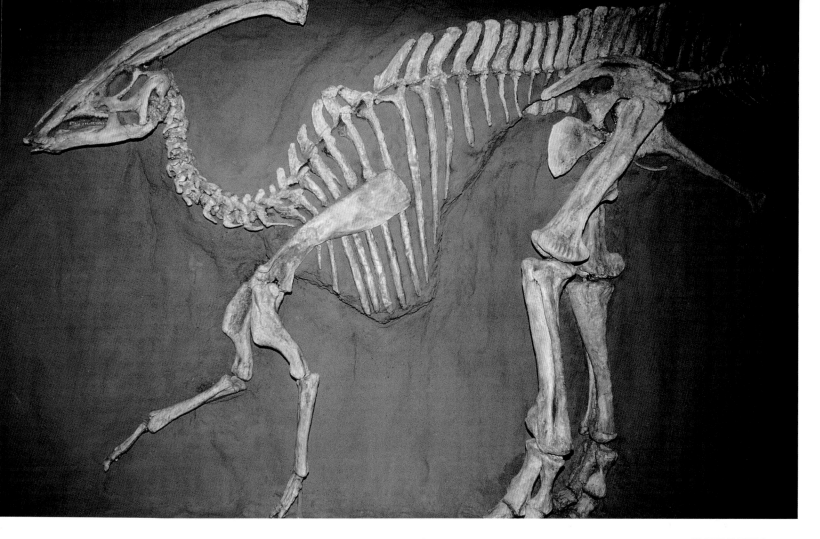

↑ **Parasaurolophus** ran on two legs, dropping to all fours to feed. It had a ducklike beak and a long, hollow crest, which it could use to trumpet.

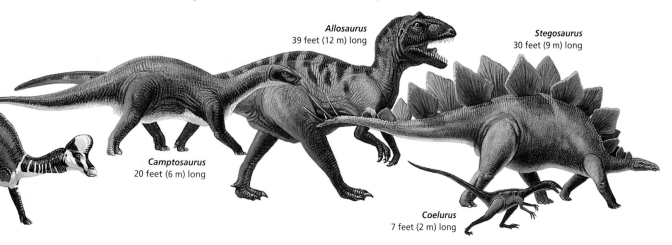

Allosaurus
39 feet (12 m) long

Stegosaurus
30 feet (9 m) long

Camptosaurus
20 feet (6 m) long

Coelurus
7 feet (2 m) long

PLANT EATERS
The warm, moist climate during the dinosaurs' reign caused vegetation to flourish. Successful planteaters had very long necks; cheek pouches to store food; stomach stones to crush tough plants or seeds; or horny beaks and specialized teeth for heavy-duty grinding.

Dinosaur tracks and traces

Dinosaurs left footprints or tracks in soft silt or mud that sometimes fossilized, leaving clues to their behavior. There are also fossils of toothmarks, nesting structures, eggs, rounded stomach stones and dung. These traces reveal much about how dinosaurs lived. Nesting sites such as Egg Mountain in Montana, USA, have shown that *Maiasaura* dinosaurs gathered and built nests in rookeries to rear their young. Their eggs were small compared to the adult size so, like birds, parents would have brought food back to satisfy the voracious appetites of their growing young. Trackway sites reveal that dinosaurs traveled and lived in groups, displaying herd instincts such as corralling juveniles in the center of a group to protect them from predators. Polished rounded stones within the ribcage of various skeletons show that some dinosaurs swallowed stones, as do some modern birds, to help grind tough food such as seeds as an aid to digestion.

WHAT DINOSAURS ATE

Toothmarks in fossil bone can indicate which animals fed on which. Late Cretaceous *Triceratops* bones with *Tyrannosaurus* toothmarks have been found in Montana, USA. Further detective work is required to determine if *Tyrannosaurus* actually hunted and killed *Triceratops* or just scavenged off the corpses. More than 500 footprints of many types of dinosaurs found at various localities reveal that huge herds once moved seasonally north and south, much like the migratory herd animals of Africa. Coprolites, or fossilized dung, can reveal much about the diet and digestive processes of dinosaurs. Hard, indigestible remains may sometimes be identified. Most dinosaurs ate plants.

↓ **Trackways tell the story** of migrating herds of longnecked sauropods moving across the North American landscape in search of food and water. They are here being pursued by predators awaiting the opportunity to take advantage of any smaller, sick or wounded animals.

← **Footprints in sandstone** need investigative work to reveal their story, such as knowing how old they are, which species made them and what they were doing. The smaller print could be of a juvenile or prey, or the smaller front foot of the same animal.

→ **Three-toed footprints** made by migrating iguanodontid dinosaurs are preserved in the sandstone of Clayton Lake State Park, USA.

Ceratopsian trackmarks show that this horn-faced, plant-eating dinosaur had a slow, four-legged gait.

Theropod trackmarks, birdlike and delicate, show this meat eater had a fast-moving, two-legged gait.

Sauropod trackmarks reveal this herbivore had a four-legged gait. Detail is of the larger, rear foot.

Pterosaurs and birds

The first reptiles took to the air in the Triassic some 220 million years ago, possibly to feast on the winged insects that had been hovering well out of reach of ground dwellers for the previous 90 million years. The pterosaurs had a thin but tough membrane joined to their bodies. These stretched between the tops of the legs and the tips of their elongated fourth fingers. Recent fossil evidence suggests that they were covered with fur and featherlike fibers. Pterosaurs were probably agile, warm-blooded fliers, capable of powered flight, rather than just gliding. The largest, *Quetzalcoatlus*, had a wingspan of more than 40 feet (12 m). Feathered *Archaeopteryx* probably evolved from the small meat-eating theropod dinosaurs 150 million years ago and likely led into the line of modern birds. Birds survived the Cretaceous mass extinction, which led to the demise of the pterosaurs 65 million years ago.

THE MISSING LINK

In 1861, an important linking fossil was discovered in the 150 million-year-old Upper Jurassic limestones at Solnhofen, Germany. *Archaeopteryx* was a feathered dinosaur, with both reptilian and birdlike features. Birds typically have feathers, opposable big toes and fused clavicles (wishbones). Reptiles typically have toothed jaws (instead of a horny beak), long bony tails, claws, and unfused wrist, hand and foot bones. *Archaeopteryx's* skull shape and rear point of attachment of the neck are also typically reptilian (birds' necks connect underneath the skull). New discoveries of less-birdlike, flightless, feathered dinosaurs in Liaoning, China, continue to add pieces to the evolutionary puzzle of powered flight.

→ **Archaeopteryx** lived during the Jurassic, 150 million years ago. It is a vitally important evolutionary link between dinosaurs and birds.

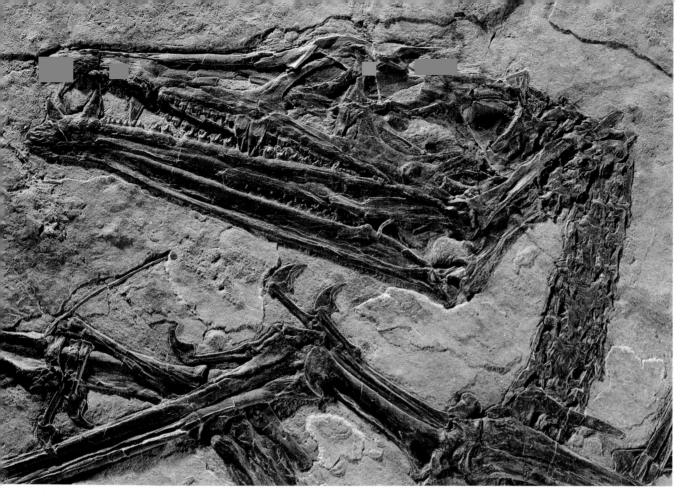

↓ **The evolution** of powered flight required major skeletal modifications, seen in this evolutionary sequence. Many scientists believe that *Archaeopteryx* was, at the very least, able to glide efficiently.

↑ **Eudimorphodon ranzii** is one of the oldest known pterosaur fossils (220 million years old). This was the first vertebrate to take to the skies.

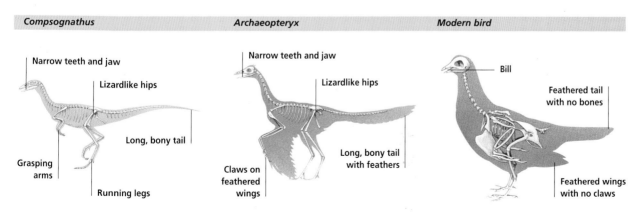

| *Compsognathus* | *Archaeopteryx* | *Modern bird* |

Narrow teeth and jaw
Lizardlike hips
Long, bony tail
Grasping arms
Running legs

Narrow teeth and jaw
Lizardlike hips
Long, bony tail with feathers
Claws on feathered wings

Bill
Feathered tail with no bones
Feathered wings with no claws

Compsognathus is a small, carnivorous dinosaur that ran upright on long, slim legs.

Archaeopteryx has features common to both dinosaurs and modern birds.

A modern bird has no tail bones, a toothless beak and a wishbone (fused clavicles).

Birds

T
K
J
Tr
P
C
D
S
O
Є
PЄ

Mammals

Mammals evolved slowly from the reptiles by developing a number of improvements, such as fur, warm blood, more sophisticated ears and efficient cuspate teeth, differentiated into canines, incisors and molars. Monotremes, the first true mammals, appeared in the early Cretaceous. They laid eggs and fed their young milk from modified sweat glands. Living egg-layers are the platypus and echidnas, found only in Australasia. Later mammals—marsupials, such as kangaroos—gave birth to tiny, live young, often keeping them in pouches. Placental mammals, including humans, now dominate the planet.

→ **Early mammal-like reptiles,** 230 million-year old herbivorous *Diictodon,* huddled together in their collapsed burrow. They predated the true mammals.

↓ **A small, hoofed, plant-eating mammal** from South America became extinct when more aggressive mammals from the north crossed the land bridge that emerged during the late Tertiary.

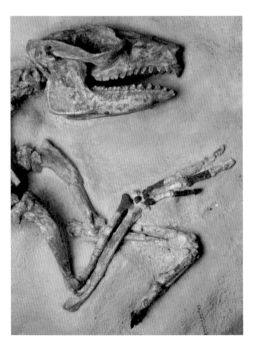

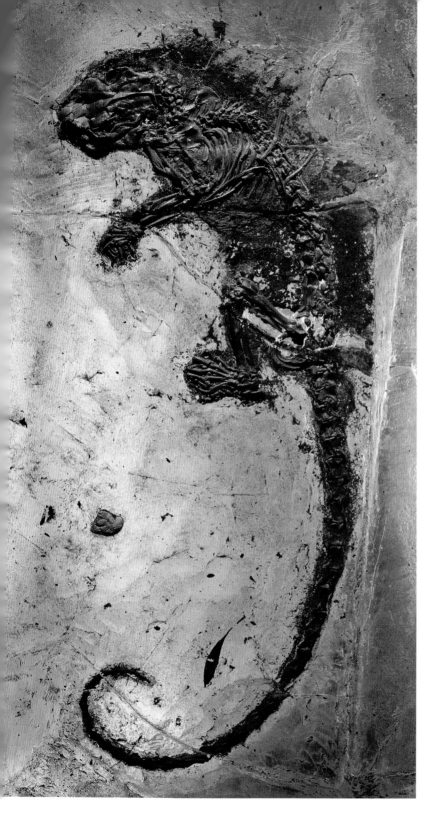

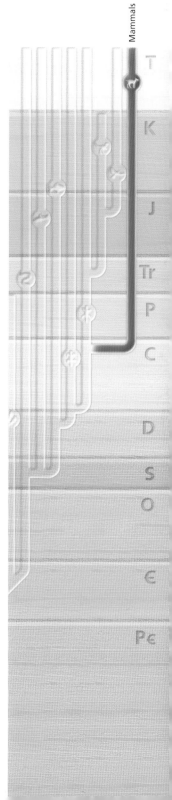

PLACENTAL MAMMALS

Placental mammals are the most successful, and they constitute 80 percent of today's 5000 known species. Their young are born in a relatively advanced state. It was only in Australia, and for a lengthy period in South America, that placental mammals did not become dominant. In many cases, such as that of the rat, there was a rapid adaptation and spreading into suitable new habitats, with the competing natives often being eliminated. Rodents are the most common placental mammals, having adapted to life in nearly every part of the globe. Another very successful placental mammal, the human, played a major part in mixing geographically separate species through exploration and trade.

↑ **The golden-mantled** ground squirrel, found along the east coast of America, reveals how its fur-covered, warm-blooded ancient counterpart (*left*) may have appeared.

← ***Ailuravus macrurus*** is a squirrel-like rodent that ate leaves and seeds. It lived 49 million years ago. The fossilized skeleton of this placental mammal was found in the oil shales of an open-pit mine near Darmstadt, Germany.

Humans

New discoveries continually modify ideas about the evolution of humans. A debate rages among paleoanthropologists over exactly which fossils are a direct ancestral line and which are merely distant cousins. It is generally agreed that evolution to the modern form involved several steps. These include an increase in height, development of bipedal movement and a change in posture from stooped to upright. Tooth and jaw size reduced, perhaps in response to changing diet. Body hair was lost. Brain size greatly increased, along with changes to the shape and size of the skull. Speech, a complex form of communication, developed. The earliest known fossil believed to be an ancestor of humanity is *Procunsul*, a Miocene hominoid. *Morotopithecus*, also Miocene, is also considered important. *Dryopithecus* lived 20 million years ago and had apelike features. Bipedal *Australopithecus* appeared 4.4 million years ago, and the genus *Homo*, roughly 2 million years ago. The oldest fossil evidence for *Homo sapiens*, the anatomically modern human, dates back at least 130,000 years.

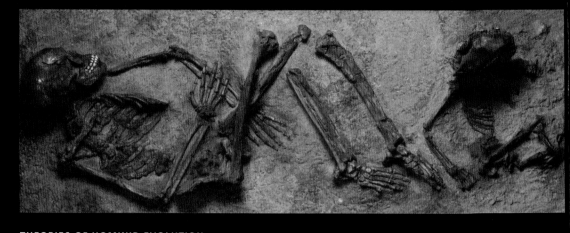

THEORIES OF HOMINID EVOLUTION

Hominids are members of the family of humans, and include the genera *Australopithecus* and *Homo*. *Homo* is thought to have arisen some 2 million years ago from *Australopithecus*. The recent discovery of *Sahelanthropus tchadensis* in Chad, central Africa, however, now puts the oldest known hominid back to 7 to 6 million years ago. Analysis suggests that it may have closer affinities with *Homo* than with *Australopithecus*. As this find predates the earliest *Australopithecus* by some 2 million years, it raises questions about our relationship to *Australopithecus*. There are also differing theories on the spread of humans. The single- origin theory contends that *Homo sapiens* emerged from Africa around 200,000 years ago. The multiregional theory proposes that *Homo erectus* left Africa about 1 million years ago and evolved into *Homo sapiens*.

↑ **These fossilized bones** of a mother and child were found at Qafzeh cave, a ritual burial site at Nazareth, Israel. These fine examples of *Homo sapiens,* the earliest anatomically modern humans, are thought to be around 90,000 to 100,000 years old.

Primitive hominid skulls, from left, show a general increase in size and development of modern features. The cranium became high-domed to take a larger brain and the brow ridges reduced in size.

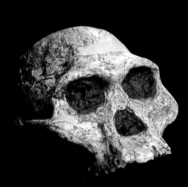

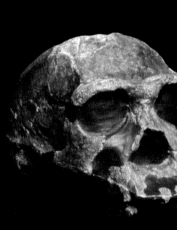

Adapis
50 million years ago

Proconsul
23 to 15 million years ago

Australopithecus africanus
3.0 to 1.8 million years ago

Homo habilis (H. rudolfensis)
2.1 to 1.6 million years ago

Homo erectus (H. ergaster)
1.8 to 0.3 million years ago

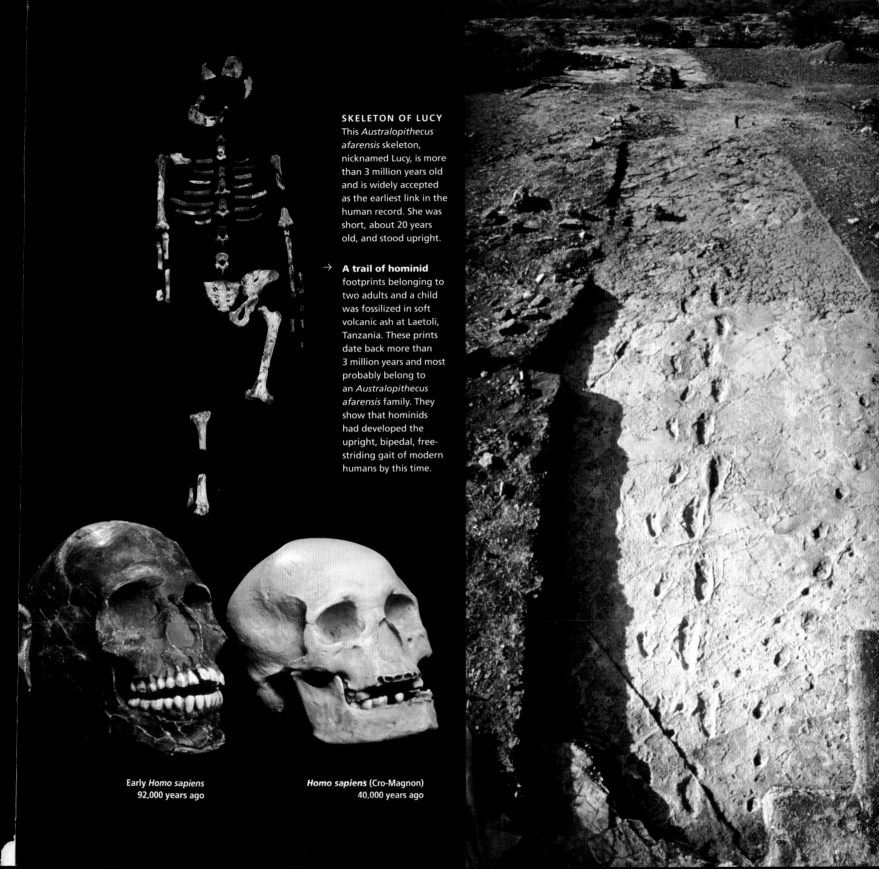

SKELETON OF LUCY
This *Australopithecus afarensis* skeleton, nicknamed Lucy, is more than 3 million years old and is widely accepted as the earliest link in the human record. She was short, about 20 years old, and stood upright.

→ **A trail of hominid** footprints belonging to two adults and a child was fossilized in soft volcanic ash at Laetoli, Tanzania. These prints date back more than 3 million years and most probably belong to an *Australopithecus afarensis* family. They show that hominids had developed the upright, bipedal, free-striding gait of modern humans by this time.

Early *Homo sapiens*
92,000 years ago

Homo sapiens (Cro-Magnon)
40,000 years ago

Factfile

NOTES

ROCKS AND MINERALS TABLES

Rock and mineral descriptions and properties are often listed in reference tables to enable their easy identification. Sometimes these tables are presented as beautifully illustrated books with one entry per page. Tables may be organized alphabetically, as here, or by particular properties. They also include many interesting details about where particular rocks and minerals can be found and their historical and present uses. The tables printed in this factfile list most of the rocks and minerals mentioned in this book and briefly summarize some of their important properties and uses. It is instructive to note just how many of the rocks are used by humans for various purposes.

IGNEOUS ROCK TABLES

The silica content of igneous rocks is dependent on the main rock-forming minerals present. For example, rocks containing quartz and feldspar are more silica-rich (silicic or acid) than those containing olivine and pyroxene, which are silica-poor (basic). The main constituent minerals also affect their appearance—silicic rocks are generally lighter in color while basic rocks are darker. Rock type and grain size clearly depend on how quickly the rock cools, either slowly underground (intrusive) or rapidly when erupted onto the surface (extrusive). Tectonic setting is important as particular igneous rocks occur only in certain settings, such as mid-oceanic ridges, convergent margins, hot spots and continental cratons.

SEDIMENTARY ROCK TABLES

Sedimentary rocks are classified according to the following criteria. They are broadly divided as either being made up of cemented fragments (clastic); the remains of plants or animals (organogenetic); or deposited from solution (chemical). The size and shape of the fragments making up the clastic rocks reveal much about the rock's depositional environment. Coarse grain size indicates a high-energy environment such as a river or violent volcanic explosion, while a fine grain size indicates deposition in a lake or in the sea, far from land. The grain shape depends on how far the fragments have been transported or reworked. The farther they have traveled, the rounder they become. Common and chemically stable minerals such as quartz are usually the main constituents of sedimentary rocks. Sedimentary rocks also tend to be associated with particular tectonic settings.

IGNEOUS ROCK TABLE

Rock	Silica content	Type	Grain size	Main constituent minerals
Andesite	intermediate	extrusive	fine	plagioclase (andesine–labradorite) biotite
Anorthosite	basic	intrusive	coarse	plagioclase (labradorite–bytownite)
Basalt	basic	extrusive	fine	plagioclase (labradorite–bytownite), augite
Dacite	intermediate	extrusive	crystals in fine matrix	plagioclase, quartz, biotite
Diorite	intermediate	intrusive	medium to coarse,	plagioclase, hornblende
Gabbro	basic	intrusive	coarse	plagioclase, olivine, clinopyroxene
Granite	acid/silicic	intrusive	coarse	quartz, orthoclase, plagioclase, mica
Granodiorite	intermediate	intrusive	coarse	quartz, plagioclase orthoclase, biotite, hornblende
Hornblendite	ultrabasic	intrusive	coarse	hornblende
Kimberlite	ultrabasic	intrusive	variable	altered olivine, phlogopite, pyrope, pyroxene
Obsidian	acid/silicic	extrusive	glassy	silica rich glass
Pegmatite	acid/silicic	intrusive	extremely coarse	quartz, orthoclase, mica often with gem minerals
Peridotite	ultrabasic	intrusive	coarse to very coarse	olivine, clinopyroxene, orthopyroxene
Pumice	acid/silicic	extrusive	glassy, vesicular	silica rich glass
Pyroxenite	ultrabasic	intrusive	coarse	clinopyroxene, orthopyroxene
Quartz porphyry	acid/silicic	extrusive	crystals in fine matrix	quartz, alkali feldspar, biotite
Rhyolite	acid/silicic	extrusive	fine	quartz, alkali feldspar
Syenite	basic	intrusive	coarse	alkali feldspar, feldspathoids
Trachyte	intermediate	extrusive	fine	sanidine, plagioclase, biotite

SEDIMENTARY ROCK TABLE

Rock	Class	Grain size	Grain shape	Main constituent minerals
Banded ironstone	organogenetic	very fine	polygonal	magnetite, hematite, chert, jasper
Breccia	clastic	variable	angular	quartz, feldspar, rock fragments
Arkose	clastic	medium	angular	feldspar, quartz, mica, rock fragments.
Coal	organogenetic	fine	fossils	plant remains: leaves, stems, trunks
Conglomerate	clastic	coarse	rounded	quartz, rock fragments
Diatomite	organogenetic	very fine	fossils	siliceous skeletal remains of diatoms etc.
Evaporite	chemical	variable	crystalline	halite, gypsum and other soluble salts
Greywacke	clastic	medium	angular	quartz, feldspar, rock fragments
Ignimbrite	pyroclastic	variable	angular	quartz, alkali feldspar, biotite and glass
Jasper	chemical	not visible		chalcedony and quartz with hematite
Limestone	organogenetic	fine to coarse	fossils	fossil fragments and calcite grains
Manganese nodules	chemical	not visible		manganese and iron oxides and hydroxides
Marl	organoclastic	very fine	variable	calcium carbonate and clay minerals
Mudstone	clastic	very fine	variable	clays and quartz
Ophicalcite	clastic	variable	angular	serpentine, gabbro cemented by calcite
Phosphorite	organogenetic	not visible		apatite, limonite, calcite and clays
Radiolarite (flint)	organogenetic	very fine	fossils	siliceous shells of radiolarians, chalcedony
Sandstone (arenite)	clastic	medium	rounded	mainly quartz, rock and mineral fragments
Shale	clastic	very fine	variable	clay, quartz, platy mica
Siltstone	clastic	fine	variable	quartz, feldspar, rock fragments and clay
Travertine	chemical	not visible		calcite or aragonite with limonite impurities
Tuff	pyroclastic	fine to variable	angular	ash (glass), augite, feldspar, olivine

Rock	Tectonic setting (where they are found)	Appearance	Uses and associations
Andesite	lava flows and domes in stratovolcanoes	blackish to greenish brown with plagioclase and biotite crystals	building, often with copper minerals
Anorthosite	differentiated ultrabasic plutons	white to gray with elongated tabular labradorite	ornamental, polished slabs, building
Basalt	mid-oceanic ridges, hot-spot volcanoes	dark to black matrix, sometimes with larger crystals	crushed for road base and concrete
Dacite	lava flows associated with convergent plates	medium gray matrix with large crystals (porphyritic)	crushed for road base and concrete
Diorite	transitional between gabbro and granodiorite	black-and-white mineral crystals, speckled appearance	building and polished slabs
Gabbro	deeper parts of large intrusions	black, dark green and minor white crystals, sometimes banded	building stone, olivine source
Granite	continental shield areas and subduction zones	light white to pink and minor black interlocking crystals	building and polished slabs
Granodiorite	plutons associated with subduction zones	white gray and lesser black mineral interlocking crystals	building and polished slabs
Hornblendite	lenses inside peridotite or around gabbro	dark green to black, coarse interlocking crystals	construction, polished slabs
Kimberlite	pipelike bodies intruding continental cratons	brecciated black, greenish to yellow when altered	associated with diamonds
Obsidian	extrusive equivalent of granite	glassy black or brown, may show banding and iridescent sheen	ornamental, carving, ancient tools
Pegmatite	veins and sheets in or around granite intrusions	white or gray, often with enormous gem crystals (topaz, beryl etc)	source of industrial minerals and metals
Peridotite	main constituent of Earth's mantle	light green, dark green, speckled black, sandy appearance	collectors item, olivine source
Pumice	gas-rich phases of plate margin volcanoes	light gray to red, abundant gas cavities allow it to float on water	abrasives, insulation
Pyroxenite	differentiated cumulate ultrabasic sequences	dark green, brown, black with elongate crystal orientation	limited use as building stone
Quartz porphyry	around plate margin stratovolcanoes	gray, pink, red matrix with large (porphyritic) crystals of quartz	building, associated with copper
Rhyolite	rapid cooling of viscous granitic rock	light colored, pink, gray, white, often flow banded with cavities	building stone, agate-filled cavities
Syenite	associated with alkaline rocks	light gray, pink to greenish, variable grain size	associated with rare metals, building
Trachyte	dikes and flows associated with hot spots	white to light gray matrix, speckled by large porphyritic crystals	building, paving and flooring

Rock	Depositional setting	Appearance	Uses
Banded ironstone	lake and marine deposits away from land	banded red (jasper) and gray to black (iron oxides)	major iron source, gemstones
Breccia	explosive volcanic deposits, talus debris	variable color, poorly sorted, angular fragments in a finer matrix	building stone, facing material
Arkose	breakdown of granite and gneiss rocks	gray, pink or reddish with poor stratification and sorting	building and paving stone
Coal	low-energy swamps, marshes and bogs	brown peat and brown coal, black bituminous to anthracite, jet	major energy source, steelmaking
Conglomerate	high-energy fluvial deposits, nearshore deltas	variable color, rounded pebbles well sorted or in a finer matrix	building stone, decorative
Diatomite	lake and marine deposits away from land	white, yellowish gray, light and porous, often friable and soft	crushed and used in filters
Evaporite	evaporite lakes, deserts, rift valley basins	white, gray, red brown, massive to earthy and crumbly	table salt, plaster, paper, alabaster
Greywacke	sediment bodies near plate collision margins	dark gray to brown, poorly size sorted angular fragments	construction stone
Ignimbrite	near plate margin stratovolcanoes	light gray or brownish to reddish, lineation of pumice lenticles	building stone
Jasper	precipitation by volcanic solutions, deep sea	gray, red, green, brown, black, massive, sub-conchoidal fracture	decorative stone, ornamental
Limestone	shallow marine distant from land sediments	very variable color, compact and massive to fossiliferous	lime, cement, ornamental, building
Manganese nodules	deep marine, distant from land	brown or black concretionary nodules, colloform banded	potential manganese source
Marl	shallow marine, lake environments	light gray, brownish to greenish, massive uniform appearance	building stone
Mudstone	offshore marine, lake environments	dark gray, black, brown, red, uniform massive appearance	none
Ophicalcite	landslides and friction/fault breccias	dark green, red, violet fragments in a white cement	attractive facing material
Phosphorite	accumulation of guano (droppings) or bones	yellowish to reddish brown, porous and usually massive	phosphatic fertilizer production
Radiolarite (flint)	lake and marine deposits away from land	white, gray, red, black, sometimes zoned, conchoidal fracture	decorative stone, cabochons
Sandstone (arenite)	medium-energy fluvial and delta deposits	variable color, often quartz rich, sandy appearance	building and paving stone
Shale	low-energy lake bed or offshore marine	variable color, splits along bedding planes, often with fossils	brickmaking
Siltstone	low-energy delta, nearshore marine	uniform brown, gray, reddish, often layered with fossils	paving stones
Travertine	around thermal spring deposits, caves	light yellowish or pinkish, finely banded, concretionary texture	construction stone
Tuff	near plate margin explosive stratovolcanoes	white, gray, brown, with larger crystals, granular, friable	light construction stone, carving

METAMORPHIC ROCK TABLE

Metamorphic rocks are broadly classified into two types—regional and contact—which reflects the broad tectonic setting in which they are formed. Regional metamorphic rocks are formed by pressure, such as that found along collision margins. Contact metamorphic rocks are formed by heating, such as experienced by rocks adjacent to large, hot, intrusive, igneous bodies. New minerals grow in the solid rock under these different conditions, and by the appearance of the rock it is immediately obvious which conditions prevailed. The minerals in contact metamorphic rocks are equigranular, such as seen in hornfels and skarn, while minerals in regionally metamorphosed rocks mostly display a preferred orientation or fabric. This is caused by the alignment of platy minerals, in particular the micas, such as seen in schist and gneiss. They grow with their flat faces at right-angles to the major regional compressive stress. Such stresses are typical of those found in continent-to-continent collision situations, such as is occurring in today's Himalayan mountains.

Metamorphic grade, being defined as either low, medium or high, reflects the amount of heat and pressure to which a rock has been subjected, up until the point when it actually starts to melt. Different minerals appear at different metamorphic grades. Some minerals prefer high pressure, or temperature, or both. A special type of high-pressure but low-temperature metamorphic rock (blueschist, eclogite) is formed when a cold oceanic plate is pushed beneath another (usually continental) plate. Certain parent rocks will produce certain metamorphic rocks under set temperature and pressure conditions. Rock chemistry reflects the chemistry or silica content of the parent rock, being either basic (for example, blueschist and eclogite from basalt), silicic (gneiss from granite), pelitic (hornfels from shale) or calcareous (marble from limestone).

USEFUL ROCKS

Important uses of the different rock types are shown in the last column of the tables, and it is interesting to note just how many have some useful purpose, such as an iron ore (ironstone) or energy source (coal). Many are used for ornamental or building purposes. In fact, any rock that is hard enough to be worked or can take a high polish—and is attractive—has been used by some society as a gemstone or decorative material; for example, coal (jet), banded ironstone, lapis lazuli, marble, granite, obsidian, limestone and serpentine. Even very soft materials, such as talc schist, pumice, travertine and tuff, have been used for carving. Other rock types are used for concrete and road aggregates.

METAMORPHIC ROCK TABLE

Rock	Type	Grade	Chemistry	Main constituent minerals
Amphibolite	regional	high	basic	amphibole, plagioclase
Blueschist	regional	low	basic	glaucophane, lawsonite
Calc schist	regional	low/med	calcareous	calcite, mica, chlorite, quartz
Eclogite	regional	medium	basic	omphacite, pyroxene, garnet
Gneiss	regional	med/high	acid/silicic	potassium feldspar, plagioclase, mica
Granulite	regional	high	acid to basic	feldspar, quartz, garnet, pyroxene
Hornfels	contact	medium	pelitic	orthoclase, plagioclase, aluminosilicates
Lapis lazuli	contact	low	calcareous	lazurite, wollastonite, calcite, pyrite
Marble	all	low to high	calcareous	calcite
Migmatite	regional	very high	acid/silicic	quartz, feldspar, biotite, hornblende
Mylonite	cataclastic	low	acid to basic	varies with surrounding rocks
Phyllite	regional	low	pelitic	quartz, sericite, mica, chlorite
Quartzite	regional	low to high	acid/silicic	quartz
Skarn	contact	high	calcareous	calcite, wollastonite, andradite, sulfides
Schist	regional	med/high	pelitic	quartz, biotite, muscovite
Serpentinite	regional	low	ultrabasic	serpentine, magnetite
Slate	regional	low	pelitic	mica, cordierite, andalusite
Talc schist	regional	low	ultrabasic	talc, with accessory dolomite, calcite

MINERAL PROPERTIES TABLE

Some of a mineral's properties are highly unique and easily measurable using simple instruments. When used in conjunction with one another, these tests can provide an unequivocal identification. Tests are considered as being destructive if they actually involve damage to the mineral surface and edges, or non-destructive if the specimen remains unharmed. Most applicable to the identification of gemstones or beautiful mineral specimens are, of course, the non-destructive tests. Certain tests may be destructive to some minerals but not others; for example specific gravity should not be carried out on water-soluble minerals. Presented here are some of the most important quantitative tests mentioned in this book.

SPECIFIC GRAVITY

Specific gravity is a most reliable and important mineral property. In practice, it can be tested with scales by measuring the normal weight of the mineral in air and comparing it with the same mineral's weight when suspended in a beaker of water. This gives the mineral's density (its weight compared to the weight of an equal volume of water). It can only be carried out on pure mineral fragments and not on minerals in a rock or gemstones set into jewelry. With a little practice it is possible to estimate specific gravity of larger specimens by hand, distinguishing between light minerals (specific gravity 1–2), such as sulfur and graphite; medium minerals (2–3), such as gypsum and quartz; medium heavy minerals (3–4), such as fluorite and beryl; heavy minerals (4–6), such as corundum and most metal oxides and sulfides; and very heavy minerals (greater than 6), such as cassiterite. Native gold and platinum are the heaviest minerals, with specific gravity of around 19.

REFRACTIVE INDEX

Refractive index and birefringence are optical properties that can be easily tested using a simple, inexpensive instrument known as a refractometer. The test can be carried out on any mineral or gemstone with a flat polished surface. It is particularly useful for gemstones set in jewelry as long as a clean, flat face can be placed against the instrument's glass measuring surface. The refractometer measures how much a light ray is bent (refracted) when passing from the instrument into the gemstone. Several measurements are taken as the stone is rotated through 360 degrees on its flat face. These results, which can be presented graphically, vary depending on the optical density, orientation and crystal system of the gemstone being tested. This test can often provide a unique gem identification, especially when used in conjunction with specific gravity. Highly skilled gemologists can make an identification by eye using a procedure known as "visual optics." In a dark room, a bright light source is viewed through a faceted gemstone being held extremely close to one eye. The position and color spread of the observed ring of sparkling rainbow patterns give a qualitative estimate of the stone's refractive index, birefringence and dispersion, and hence shed light on the identity of the gemstone.

Rock	Original source rocks	Appearance	Uses and associations
Amphibolite	basic igneous rocks, basalt, dolerite, gabbro	dark green to black, coarse grain size, massive to banded	construction, polished slabs
Blueschist	basalt, basaltic tuff, gabbro, subduction zones	pale blue to violet, often with relict parent rock structure	rare, attractive rock type
Calc schist	calcareous shale or calcareous arenite	gray to brown gray, alternating granular and lamellar units	building material
Eclogite	basalt, basaltic tuff, gabbro, subduction zones	variable grain size, green to red, massive, spotted	attractive collectors' specimen
Gneiss	felsic sedimentary rock, granite, granodiorite	coarse-grained light to dark, foliated with schistose mica	building material, rough or polished
Granulite	lower crustal rocks, Precambrian shield areas	variable light to dark, massive spotted with garnets	building stone and polished slabs
Hornfels	clays, shales, mudstones	fine-grained, gray to brown, non-laminated, cherty looking	road base and construction
Lapis lazuli	impure limestones, dolomitic evaporites	fine-grained blue with spots or patches of white and gold	gemstone and ornamental, historic paint
Marble	pure calcareous limestone	fine to coarse, pure white to patchy gray, brown and red	building, polished slabs, sculpture
Migmatite	partial melting of a granitic or pelitic rock	white to dark gray, foliated gneissic to granitic texture	building stone and polished slabs
Mylonite	rocks ground and brecciated in a fault zone	often brecciated or highly foliated appearance	construction gravel
Phyllite	clayey to sandy sedimentary rocks	fine-grained, silvery to greenish, strong micaceous foliation	rarely used as roofing slabs
Quartzite	quartz-rich sedimentary rocks	white to patchy gray, sandy or massive appearance	construction and flooring material
Skarn	impure limestone adjacent to granite plutons	coarse, variable color, minerals radiating or banded	building, copper, molybdenum, iron
Schist	clayey to sandy to limey sedimentary rocks	silvery to shiny black, strong micaceous foliation	decorative uses
Serpentinite	ultrabasic peridotite, pyroxenite, amphibolite	pale to dark green, often slickensided due to shearing	building, ornamental, chrome, nickel
Slate	clays and shales	fine-grained, shiny, dark gray, micaceous laminations	roofing, flooring slabs, blackboards
Talc schist	olivine bearing ultrabasic rocks	grayish white to speckled green, greasy feel, very soft	industrial talc, cosmetics (talcum)

CHEMICAL FORMULA

Formula and chemical group can be obtained analytically by using a small fragment of the mineral, which must be either polished or powdered. The formula is presented as symbols; for example, Si for silicon and O for oxygen. Numbers indicate how many atoms of each there are—quartz SiO_2 has two oxygen atoms for each silicon. While it is the most definitive way of identifying a mineral, this test is not routinely used in gem identification because of the time and expense involved.

HARDNESS

Hardness is often used by geologists in the field using a set of reference minerals. It is a destructive test, as a clear scratch must be made in the mineral's surface by one of the minerals in the reference set to determine its relative position on Mohs' hardness scale. For this reason, gemologists never use the hardness test during the course of carrying out routine identifications on crystals and gemstones in the laboratory.

COLOR

Color as a means of mineral identification must be used with extreme caution, especially by beginners, as many minerals are allochromatic, such as fluorite, and display a confusing variety of colors. A few minerals have a unique (idiochromatic) color and pattern, which often allows them be recognized by this property without further testing. Examples include malachite (green) and rhodochrosite (pink), with their unique banding; and rhodonite, with its distinctive black manganese dendrites.

CRYSTAL SYSTEM

Crystal system and habit both influence the common appearance of a mineral, and as such these may often greatly aid in hand identification of an unknown sample. Consider the unmistakable crystal shape of fine specimens of quartz, beryl, garnet, pyrite and diamond. Or consider the diagnostic habit of the micas (platy), chrysotile asbestos (fibrous), malachite (botryoidal) and rhodochrosite (reniform, stalactitic), native mercury (liquid) and natrolite (acicular, stellate).

STREAK

Streak is a destructive test, as an edge of the specimen must be rubbed along an unglazed porcelain tile. For most minerals it is just white. However, for a few minerals, streak can be quite diagnostic. Consider pyrite, a gold-colored mineral that produces a greenish black streak; hematite, a gray mineral that produces a cherry red streak; and cassiterite, a black mineral that produces a yellow or white streak.

LUSTER

Luster is the appearance of light reflected off a mineral's surface. It may vary depending on whether the surface is fresh or tarnished, or if the specimen is a single crystal or a fine crystal aggregate. It is especially useful when used in conjunction with sheen and color. Most of the metal sulfides and native metals have a distinctive metallic luster. Diamond has quite a unique adamantine luster. Minerals like turquoise and talc have a greasy luster, and kaolinite is just plain dull.

SHEEN

Sheen is the resultant optical effect caused by light reflecting from a mineral's internal structure or inclusions. It can be uniquely diagnostic. Consider the rainbow play of color displayed by precious opal, the iridescence of labradorite feldspar and the blue schiller of moonstone feldspar. Fine hairlike inclusions are responsible for the cat's eye chatoyancy of chrysoberyl and the star asterism of corundum. The sparkle in sunstone and aventurine quartz is also due to tiny inclusions.

GEMSTONE MINERALS

Gemstone or ornamental applications of the various minerals are surprisingly common, even some that one wouldn't imagine, such as hematite, the principal ore of iron. The shiny, silver-gray colored jewelry made from hematite is attractive, even if a little heavy. If a mineral is rare and yet beautiful, then it becomes one of humankind's truly precious gems. These include diamonds, sapphires, rubies, emeralds and opals.

TECTONIC SETTING

Knowledge of the tectonic setting of a mineral's host rock can narrow down the identification choices. Lazurite, constituent of the rock lapis lazuli, is found only in a few global localities in metamorphosed limestone or dolomite. Furthermore, when worked, it produces an unpleasant smell due to sulfur in its chemical composition. Green and blue minerals found near a copper mine are likely to contain copper as a component, thereby limiting the possibilities of what they could be.

MINERAL PROPERTIES TABLE

Mineral	Formula	Mineral group	Hardness	Specific gravity	Common color	Crystal system
Agate	SiO_2	oxide	6.5–7	2.5–2.61	various colors, banded	trigonal
Albite	$NaAlSi_3O_8$	silicate	6–6.5	2.61	gray, white, bluish	triclinic
Almandine	$Fe_3Al_2(SiO_4)_3$	silicate	6.5–7.5	3.4–4.6	purplish red	cubic
Amazonite	$KAlSi_3O_8$	silicate	6	2.5–2.6	green, blue-green	triclinic
Amethyst	SiO_2	oxide	7	2.65	purple, sometimes zoned	trigonal
Andalusite	Al_2SiO_5	silicate	7.5	3.1–3.2	gray, brown, red, green	orthorhombic
Anorthite	$CaAl_2Si_2O_8$	silicate	6–6.5	2.76	gray, white, bluish	triclinic
Apatite	$Ca_5(PO_4)_3(F,OH)$	phosphate	5	3.16–3.22	yellow, green, many colors	hexagonal
Apophyllite	$KCa_4Si_8O_{20}(F,OH).8H_2O$	silicate	4.5–5	2.3–2.4	white, red, green, violet	tetragonal
Aquamarine	$Be_3Al_2Si_6O_{18}$	silicate	7.5–8	2.65–2.8	blue green, light blue	hexagonal
Aragonite	$CaCO_3$	carbonate	3.5–4	2.95	white, yellowish, bluish	orthorhombic
Argentite	Ag_2S	sulfide	2	7.3	dark gray to black	cubic
Arsenopyrite	$FeAsS$	sulfide	5.5–6	5.9–6.2	light gray, white	monoclinic
Atacamite	$Cu_2Cl(OH)_3$	halide	3–3.5	3.76	green to dark green	orthorhombic
Augite	$(Ca,Na)(Mg,Fe,Al,Ti)(Si,Al)_2O_6$	silicate	5.5–6	3.3–3.5	dark green to black	monoclinic
Azurite	$Cu_3(CO_3)_2(OH)_2$	carbonate	3.5–4	3.7–3.9	light to dark blue	monoclinic
Barite	$BaSO_4$	sulfate	3–3.5	4.4	colorless, white	orthorhombic
Beryl	$Be_3Al_2Si_6O_{18}$	silicate	7.5–8	2.65–2.8	white, many colors	hexagonal
Biotite	$K(Mg,Fe)_3(Al,Fe)Si_3O_{10}(OH)_2$	silicate	2.5–3	2.8–3.2	black, dark brown	monoclinic
Calcite	$CaCO_3$	carbonate	3	2.6–2.8	colorless, white, yellow	trigonal
Carnallite	$KMgCl_3.6H_2O$	halide	2.5	1.6	colorless, white, pink	orthorhombic
Carnelian	SiO_2	silicate	6.5–7	2.5–2.61	translucent red	trigonal
Cassiterite	SnO_2	oxide	7	6.8–7.1	black	tetragonal
Cerussite	$PbCO_3$	carbonate	3–3.5	6.4–6.6	colorless, white, yellow	orthorhombic
Chalcanthite	$CuSO_4.5H_2O$	sulfate	2.5	2.2–2.3	blue	triclinic
Chalcedony	SiO_2	oxide	6-7	2.5–2.61	many colors, color bands	trigonal
Chalcopyrite	$CuFeS_2$	sulfide	3.5–4	4.2–4.3	golden yellow	tetragonal
Chrysoberyl	$BeAl_2O_4$	oxide	8.5	3.7	yellow, brown, green	orthorhombic
Chrysocolla	$(Cu,Al)_2H_2Si_2O_5(OH)_4.nH_2O$	silicate	2–4	2.0	green, blue, blue-green	monoclinic
Chrysoprase	SiO_2	oxide	6.5–7	2.5–2.61	bright green	trigonal
Chrysotile	$Mg_3Si_2O_5(OH)_4$	silicate	3–4	2.5–2.6	gray, blue, yellow, greenish	monoclinic
Cinnabar	HgS	sulfide	2–2.5	8.1	red	trigonal
Citrine	SiO_2	oxide	7	2.65	pale to golden yellow	trigonal
Copper	Cu	native element	2.5–3	8.93	light red	cubic
Cordierite	$Mg_2Al_4Si_5O_{18}$	silicate	7	2.6	blue, gray, violet	orthorhombic
Corundum	Al_2O_3	oxide	9	4.0	black, many colors	trigonal
Cuprite	Cu_2O	oxide	3.5–4	6.15	red, brown, black	cubic
Diamond	C	native element	10	3.52	colorless, many colors	cubic
Diopside	$CaMgSi_2O_6$	silicate	5–6	3.3	green to black	monoclinic
Elbaite	$Na(Li,Al)_3Al_6(BO_3)_3Si_6O_{18}(OH)_4$	silicate	7–7.5	2.9–3.2	red, green, color zoned	trigonal
Emerald	$Be_3Al_2Si_6O_{18}$	silicate	7.5–8	2.65–2.8	light to dark green	hexagonal
Enstatite	$Mg_2Si_2O_6$	silicate	5.5	3.1–3.2	dark green to black	orthorhombic
Epsomite	$MgSO_4.7H_2O$	sulfate	2–2.5	1.68	white, light pink, yellow	orthorhombic
Fluorite	CaF_2	halide	4	3.18	all colors of spectrum	cubic
Galena	PbS	sulfide	2.5	7.2–7.6	silver to lead gray	cubic
Gold	Au	native element	2.5–3	19.2	golden yellow	cubic
Graphite	C	native element	1–1.5	2.25	gray, dark gray, black	hexagonal
Grossular	$Ca_3Al_2(SiO_4)_3$	silicate	6.5–7.5	3.4–4.6	yellowish green	cubic
Gypsum	$CaSO_4.2H_2O$	sulfate	2	2.3–2.4	colorless, white, yellow	monoclinic
Halite	$NaCl$	halide	2	2.1–2.2	colorless, pale colors	cubic
Hematite	Fe_2O_3	oxide	6.5	5.2–5.3	gray, black	trigonal

Mineral	Habit	Streak	Luster	Gemstone	Tectonic setting and other notes
Agate	cavity fillings	white	vitreous/silky	ornamental	concentric often banded cavity fillings, chalcedony-type quartz
Albite	prisms, plates	white	vitreous/pearly	semiprecious	sodium plagioclase feldspar variety, common rock-forming mineral
Almandine	rhombs (soccer balls)	white	vitreous	semiprecious	pegmatites, metamorphic, common iron-rich garnet group mineral
Amazonite	short prisms, twins	white	vitreous/pearly	semiprecious	giant crystals in pegmatites, potassium feldspar, microcline variety
Amethyst	cavity filling crystals	white	vitreous	semiprecious	magmatic, hydrothermal deposits, purple crystalline quartz variety
Andalusite	columnar, radiating, fibrous	white	vitreous/greasy	semiprecious	metamorphic, pegmatites, chiastolite has inclusions forming a cross
Anorthite	prisms, plates	white	vitreous/pearly		calcium plagioclase feldspar, common rock-forming mineral
Apatite	prisms, plates	white	vitreous/greasy	semiprecious	magmatic, pegmatites, hydrothemal, component of phosphorites
Apophyllite	plates, massive	white	vitreous/pearly		hydrothermal associated with calcite, analcime, natrolite
Aquamarine	columnar prisms	white	vitreous	semiprecious	large crystals in pegmatites, common gem beryl variety
Aragonite	fibrous, stellate	white	vitreous	ornamental	hydrothermal, sedimentary, found with calcite, zeolites, limonite
Argentite	equant, massive	black	dull		secondary, hydrothermal, important silver ore, silver/lead deposits
Arsenopyrite	prisms, aggregates	black	metallic		hydrothermal, metamorphic, arsenic ore found with stibnite, galena
Atacamite	columns, plates, fibrous	green	vitreous		secondary, fumaroles, ore found with other copper minerals
Augite	short prisms	gray-green	vitreous	semiprecious	basic igneous rocks, pyroxene, common rock-forming mineral
Azurite	plates, massive	light blue	vitreous/dull	semiprecious	secondary with copper minerals, ore of copper and blue pigment
Barite	tabular, aggregates	white	vitreous/pearly	semiprecious	hydrothemal or evaporite mineral, used as drilling mud, barium ore
Beryl	columnar prisms	white	vitreous	semiprecious	large crystals in pegmatite, varieties; emerald, aquamarine, heliodor
Biotite	plates, short prisms	gray	vitreous/pearly		pegmatites, common rock-foming mineral, mica group member
Calcite	crystals, aggregates	white	vitreous	ornamental	evaporite/hydrothermal, metamorphic, also stalactites in caves
Carnallite	granular, fibrous aggregates	white	vitreous/greasy		evaporite mineral, soluble, source of potassium and magnesium
Carnelian	cavity fillings	white	vitreous/silky	ornamental	cavity fillings, variety of chalcedony colored by hematite inclusions
Cassiterite	octahedra, aggregates	yellow	vitreous	semiprecious	pegmatitic, hydrothermal, secondary, found in placers, ore of tin
Cerussite	reticulated, aggregates	white	adamantine		secondary mineral found in lead zinc deposits, ore of lead
Chalcanthite	stalactitic, reniform	white	vitreous		secondary sulfate associated with copper orebodies, soluble, toxic
Chalcedony	crusts and fillings	white	vitreous/silky	ornamental	secondary, includes agate, carnelian, chrysoprase, jasper, onyx
Chalcopyrite	compact aggregates	greenish black	metallic	semiprecious	magmatic, metamorphic, hydrothemal, most important copper ore
Chrysoberyl	short prisms, cyclic twins	white	vitreous/greasy	semiprecious	pegmatites, alexandrite changes from green by day to red at night
Chrysocolla	botryoidal, massive	light green	greasy	semiprecious	secondary mineral, in deposits with copper oxides and carbonates
Chrysoprase	cavity and vein fillings	white	vitreous/silky	semiprecious	weathered ultrabasic rocks, variety of chalcedony colored by nickel
Chrysotile	fibrous, in veins	gray	silky	semiprecious	hydrothermal alteration of ultrabasic rocks, a type of asbestos
Cinnabar	microcrystalline to earthy	red	dull		hydrothermal mineral, mercury ore, found with stibnite, marcasite
Citrine	cavity filling crystals	white	vitreous	semiprecious	magmatic, hydrothermal deposits, yellow crystalline quartz variety
Copper	arborescent, wiry	copper red	metallic	native metal	found in deposits with copper oxides/carbonates, ore of copper
Cordierite	short prisms	white	vitreous/greasy	semiprecious	metamorphic, magmatic, pegmatitic, gem variety: iolite/dichroite
Corundum	pyramidal, tabular	white	vitreous	precious	magmatic, metamorphic, in alluvials, varieties: sapphire, ruby (red)
Cuprite	equant crystals, earthy	brownish red	dull		secondary, in deposits with copper carbonates, ore of copper
Diamond	octahedra, cubes	white	adamantine	precious	carried from the mantle by ultrabasic magmas, found in alluvials
Diopside	short prisms	white	vitreous/greasy	semiprecious	basic to ultrabasic rocks, common rock-forming pyroxene mineral
Elbaite	columnar prisms	white	vitreous	semiprecious	large crystals in pegmatite, color zoned, tourmaline group mineral
Emerald	columnar prisms	white	vitreous	precious	metamorphic and pegmatitic rocks, precious gem beryl variety
Enstatite	short prisms, fibrous	white	vitreous	semiprecious	mafic to ultrabasic igneous rocks, common rock-forming pyroxene
Epsomite	fibrous, crusts, stalactitic	white	vitreous/silky		evaporite mineral, soluble, bitter saline taste, used for epsom salts
Fluorite	cubes, octahedral cleavage	white	vitreous	ornamental	hydrothermal, pegmatites, with galena and sphalerite, used as flux
Galena	cubes, octahedra	shiny gray	metallic		hydrothermal mineral, primary ore of lead, found with sphalerite
Gold	crystals, dendritic, nuggets	gold	metallic	native metal	hydrothemal, secondary mineral, heavy, found in alluvial placers
Graphite	lamellar, earthy	dark gray	metallic, earthy		metamorphic, pegmatites, uses; lubricant, pencil lead, reactors
Grossular	dodecahedra, trapezohedra	white	vitreous	semiprecious	rare contact metamorphic garnet group member, calcium garnet
Gypsum	plates, massive, earthy	white	vitreous/pearly	ornamental	common evaporite mineral; alabaster is used for carving
Halite	equant, earthy	white	vitreous/greasy		common evaporite mineral, crushed for industrial and table salt
Hematite	massive, earthy	red	dull	ornamental	magmatic, hydrothermal, sedimentary deposits, principal ore of iron

MINERAL PROPERTIES TABLE

Mineral	Formula	Mineral group	Hardness	Specific gravity	Common color	Crystal system
Jadeite	$Na(Al,Fe)Si_2O_6$	silicate	6.5	3.2–3.3	gray, light to dark green	monoclinic
Kaolinite	$Al_2Si_2O_5(OH)_4$	silicate	1	2.6	white, yellowish, brownish	triclinic
Kyanite	Al_2SiO_5	silicate	4-7	3.6–3.7	white, blue, gray	triclinic
Labradorite	$(Na,Ca)Al_{1-2}Si_{3-2}O_8$	silicate	6–6.5	2.7	gray, white, bluish	triclinic
Lazurite	$(Na,Ca)_8(Al,Si)_{12}(O,S)_{24}[(SO_4,Cl_2,(OH)_2]$	silicate	5.5	2.38–2.42	dark blue, violet blue	cubic
Lepidolite	$K(Li,Al)_3(Si,Al)_4O_{10}(OH,F)_2$	silicate	2.5–3	2.8–2.9	violet, light red	monoclinic
Magnetite	Fe_3O_4	oxide	5.5	5.2	black	cubic
Malachite	$Cu_2(CO_3)(OH)_2$	silicate	4	4.0	light to dark green, banded	monoclinic
Mercury	Hg	native element	liquid	13.6	gray white	none-liquid
Millerite	NiS	sulfide	3.5	5.3	light yellow to brown	trigonal
Molybdenite	MoS_2	sulfide	1–1.5	4.7–4.8	bluish grey	hexagonal
Moonstone	$KAlSi_3O_8$	silicate	6	2.53–2.56	colorless to bluish	monoclinic
Muscovite	$KAl_2AlSi_3O_{10}(OH)_2$	silicate	2–2.5	2.7–2.8	silver white, gray	monoclinic
Natrolite	$Na_2Al_2Si_3O_{10}.2H_2O$	silicate	5–5.5	2.2	white, gray, yellow, brown	orthorhombic
Nephrite	$Ca_2(Mg,Fe)_5Si_8O_{22}(OH)_2$	silicate	5–6	3.0–3.2	light to dark green	monoclinic
Olivine	$(Mg,Fe)_2SiO_4$	silicate	6.5–7	3.27–4.20	yellow green, olive green	orthorhombic
Opal	$SiO_2.nH_2O$	silicate	5.5–6.5	2.1–2.2	white, black, all colors	amorphous
Orthoclase	$KAlSi_3O_8$	silicate	6	2.53–2.56	colorless, white, yellow	monoclinic
Platinum	Pt	native element	4–4.5	14–19	silvery white, gray	cubic
Pyrite	FeS_2	sulfide	6–6.5	5.0–5.2	brassy yellow	cubic
Pyrolusite	MnO_2	oxide	6–7	5.0	dark gray to black	tetragonal
Pyrope	$Mg_3Al_2(SiO_4)_3$	silicate	6.5–7.5	3.4–4.6	red to violet red	cubic
Quartz	SiO_2	oxide	7	2.65	colorless, white, all colors	trigonal
Realgar	As_4S_4	sulfide	1.5	3.5	orange-red, red	monoclinic
Rhodochrosite	$MnCO_3$	carbonate	4	3.3–3.6	pink to red	trigonal
Rhodonite	$(Mn,Fe,Mg,Ca)SiO_3$	silicate	5.5–6.5	3.73	pink to red	triclinic
Ruby	Al_2O_3	oxide	9	4.0	pink to red	trigonal
Rutile	TiO_2	oxide	6–6.5	4.2–4.3	red, brown, black	tetragonal
Sapphire	Al_2O_3	oxide	9	4.0	blue, green, yellow, violet	trigonal
Sassolite	H_3BO_3	hydroxide	1	1.45	colorless, white, gray	triclinic
Scapolite	$(Na,Ca,K)_4Al_3(Al,Si)_3Si_6O_{24}(Cl,SO_4,CO_3)$	silicate	5–6	2.54–2.77	white gray, blue, pink	tetragonal
Sillimanite	Al_2SiO_5	silicate	6–7	3.2	yellow, gray, light blue	orthorhombic
Silver	Ag	native element	2.5–3	10–12	silvery white, gray	cubic
Smithsonite	$ZnCO_3$	carbonate	5	4.3–4.5	white, yellow, red, green	trigonal
Spessartine	$Mn_3Al_2(SiO_4)_3$	silicate	6.5–7.5	3.4–4.6	yellow, orange, pink	cubic
Sphalerite	$(Zn,Fe)S$	carbonate	3.5–4	3.9–4.2	black to brownish red	cubic
Spinel	$MgAl_2O_4$	oxide	8	3.5	yellow, blue, green, red	cubic
Staurolite	$Fe_2Al_9(Si,Al)_4O_{22}(OH)_2$	silicate	7–7.5	3.5–3.6	brown to black	orthorhombic
Stibnite	Sb_2S_3	sulfide	2	4.6–4.7	gray to black	orthorhombic
Sulfur	S	native element	1.5–2	2.05–2.08	yellow	monoclinic
Talc	$Mg_3Si_4O_{10}(OH)_2$	silicate	1	2.8	white, gray, green, yellow	monoclinic
Tanzanite	$Ca_2Al_3(SiO_4)_3(OH)$	silicate	6–6.5	3.2–3.4	blue, violet blue	orthorhombic
Tetrahedrite	$(Cu,Fe,Ag,Zn)_{12}Sb_4S_{13}$	sulfide	3–4.0	4.6–5.2	dark gray to black	cubic
Topaz	$Al_2SiO_4(F,OH)_2$	silicate	8	3.5–3.6	colorless, pale colors	orthorhombic
Turquoise	$CuAl_6(PO_4)_4(OH)_8.4H_2O$	phosphate	5–6	2.6–2.8	light blue, light green	triclinic
Uraninite	UO_2	oxide	6	10.6	gray to black	cubic
Vermiculite	$(Mg,Fe,Al)_3(Al,Si)_4O_{10}(OH)_2.4H_2O$	silicate	1.5	2.3–2.7	yellowish brown	monoclinic
Waveliite	$Al_3(PO_4)_2(F,OH)_3.5H_2O$	phosphate	3.5–4	2.3–2.4	white, pale green-blue	orthorhombic
Wolframite	$(Fe,Mn)WO4$	tungstate	5–5.5	7.1–7.5	dark brown to black	monoclinic
Wulfenite	$PbMoO_4$	molybdate	3	6.7–6.9	yellow, orange, reddish	tetragonal
Zircon	$ZrSiO_4$	silicate	7.5	4.0–4.7	colorless, yellow to red	tetragonal

Mineral	Habit	Streak	Luster	Gemstone	Tectonic setting and other notes
Jadeite	compact granular aggregates	white	vitreous/greasy	semiprecious	metamorphic, pyroxene group mineral commonly termed jade
Kaolinite	massive, earthy	white	dull		weathering product, hydrothermal, used for ceramics, bricks
Kyanite	columns, radial, massive	white	vitreous/pearly	semiprecious	common metamorphic, pegmatites, distinct differential hardness
Labradorite	prisms, plates, aggregates	white	vitreous/pearly	semiprecious	magmatic, plagioclase feldspar, often with spectacular blue sheen
Lazurite	compact massive	light blue	vitreous	semiprecious	contact metamorphic, with calcite and pyrite in the rock lapis lazuli
Lepidolite	plates, aggregates	light pink/white	pearly		pegmatites, hydrothermal, ore of lithium, rare mica group member
Magnetite	octahedra, massive	black	metallic, dull		magmatic, metamorphic, hydrothemal, most valuable ore of iron
Malachite	botryoidal, massive	light green	vitreous/silky	semiprecious	secondary, in deposits with other copper minerals, ore of copper
Mercury	liquid at room temperature		metallic		hydrothermal, oxidized zones, associated with cinnabar
Millerite	acicular, stellate	greenish black	metallic		hydrothemal, magmatic, sedimentary, rare ore of nickel
Molybdenite	scales, lamellar aggregates	blue-gray	metallic		magmatic, pegmatites, hydrothermal, ore of molybdenum
Moonstone	tabular, aggregates	white	vitreous, pearly	semiprecious	orthoclase feldspar variety; displays a blue-schiller sheen
Muscovite	sheets, aggregates	white	pearly, silky		pegmatites, common rock-forming mineral, mica group member
Natrolite	acicular, stellate	white	vitreous/silky		hydrothemal, associated with other zeolite minerals and calcite
Nephrite	massive aggregates	white	vitreous/silky	semiprecious	contact metamorphic, with talc and serpentine, amphibole group
Olivine	short prisms	white	vitreous	semiprecious	magmatic, main constituent of Earth's mantle, meteorites, alluvials
Opal	massive	white	waxy, greasy	precious	secondary, precious varieties show play of spectral colours
Orthoclase	prisms, aggregates	white	vitreous/pearly	semiprecious	common rock-forming potassium feldspar, uses: ceramics, glass
Platinum	nuggets	white	metallic	native metal	rare ultrabasic igneous mineral, heavy, found in alluvial deposits
Pyrite	cubes, rhombs, aggregates	greenish black	metallic	semiprecious	magmatic, metamorphic, pegmatitic, often with metal sulfide ores
Pyrolusite	dendrites, massive, botryoidal	black	submetallic, dull		secondary, hydrothermal, weathering crusts, manganese ore
Pyrope	rhombs (soccer balls)	white	vitreous	semiprecious	ultrabasic igneous rocks, garnet group member, diamond indicator
Quartz	columnar, massive	white	vitreous	semiprecious	rock-forming mineral, includes amethyst, citrine, rose and smoky
Realgar	striated prisms	orange red	adamantine		hydrothermal and hot springs, secondary, ore of arsenic
Rhodochrosite	reniform, stalactitic	white	vitreous	semiprecious	hydrothermal, contact metamorphic, pegmatites, cave deposits
Rhodonite	massive	white	vitreous	semiprecious	hydrothermal, metamorphic, associated with manganese ores
Ruby	tabular prisms	white	vitreous	precious	metamorphic, gem corundum variety, found in alluvial deposits
Rutile	striated prisms, twins	yellow brown	sub-adamantine	semiprecious	magmatic, pegmatites, metamorphic, alluvial deposits, titanium ore
Sapphire	pyramidal to columnar	white	vitreous	precious	igneous or metamorphic, gem corundum, found in alluvial deposits
Sassolite	sheets, plates	gray, white	vitreous/pearly		volcanic hot springs, water soluble, bitter taste, rare mineral
Scapolite	columnar, fibrous	white	vitreous/pearly	semiprecious	contact metamorphic, associated with garnet and pyroxene
Sillimanite	acicular (needle-like)	white	vitreous/silky	semiprecious	common contact or regional metamorphic, uses: insulator
Silver	equant crystals, wires, plates	white	metallic	native metal	hydrothermal, in deposits with lead and silver sulfides
Smithsonite	stalactitic, botryoidal, crusts	white	vitreous, pearly	semiprecious	secondary, oxidation zones of lead–zinc orebodies, ore of zinc
Spessartite	rhombs (soccer balls)	white	vitreous	semiprecious	pegmatites, metamorphic, manganese-rich garnet group member
Sphalerite	equant, massive	light brown	sub-adamantine		hydrothermal, pegmatites, principal ore of zinc, found with galena
Spinel	octahedra	white	vitreous	semiprecious	magmatic, metamorphic, spinel group member, found in alluvials
Staurolite	short prisms, twins	gray	vitreous/dull	semiprecious	metamorphic, forms cross-shaped twins, found in alluvial deposits
Stibnite	long columns, fibrous	gray	metallic		hydrothermal, important ore of antimony, found with realgar/gold
Sulfur	dipyramidal crystals, earthy	light yellow	adamantine		hot springs, sedimentary, oxidized zone of sulfide ore bodies
Talc	massive, lamellar aggregates	white	pearly, greasy		hydrothermal, contact metamorphic, uses: lubricant, paper, carving
Tanzanite	columnar prisms	white	vitreous/pearly	semiprecious	metamorphic, rare blue zoisite only from Tanzania, epidote group
Tetrahedrite	tetrahedra, massive	black	metallic		hydrothemal, associated with copper–zinc orebodies, copper ore
Topaz	columnar and short prisms	white	vitreous	semiprecious	crystals often meters long, in pegmatites and hydrothermal deposits
Turquoise	massive, veins, crusts	white	greasy	semiprecious	rare secondary mineral, associated with limonite and chalcedony
Uraninite	massive, earthy, reniform	brown-green	dull		pegmatites, hydrothermal, sedimentary, radioactive, uranium ore
Vermiculite	sheets, scaly aggregates	greenish	pearly		secondary mineral, uses: heat and sound insulation, lubricants
Wavellite	globular, radiating	white	vitreous/silky		secondary, hydrothermal, found with limonite, hematite, pyrolusite
Wolframite	prisms, plates	black	metallic/greasy		pegmatites, hydrothermal, tungsten ore, with scheelite, molybdenite
Wulfenite	plates, granular aggregates	light yellow	adamantine		secondary, ore of lead and molybdenum, found with galena, calcite
Zircon	short prisms	white	sub-adamantine	semiprecious	magmatic, metamorphic, pegmatites, found in alluvial placers

GEOLOGICAL TIMESCALE

Era	Period (million years ago)	Epoch	Evolutionary events
Cenozoic (65–present)			
	Quaternary (1.8–present)	Recent (0.01–present)	First anatomically modern human (*Homo*); migrations cause extinction of large mammals; rise of civilizations
		Pleistocene (1.8–0.01)	Global ice age begins; glaciers push out from the polar regions; mammals develop woolly coats; hominids evolve
	Tertiary (65–1.8)	Pliocene (5.3–1.8)	Grazing herbivores grow larger and prosper on grasslands; predators include saber-tooth cats, dogs and bears
		Miocene (24–5.3)	Grasses become important in colder conditions and plains animals develop; hominid–chimpanzee lines separate
		Oligocene (34–24)	Climate becomes increasingly cool; sea levels drop as ice accumulates on land and Antarctica freezes over
		Eocene (55–34)	Temperatures cool; woodlands replace forests; mammals take to air (bats) and sea (whales); primates appear
		Paleocene (65–55)	Dense forests; warm moist climate; new mammals diversify and rise to dominance, filling niches left by dinosaurs
Mesozoic (248–65)			
	Cretaceous (144–65)		Flowering plants (angiosperms) and insects diversify; gymnosperms decline; dinosaurs wiped out in KT extinction
	Jurassic (206–144)		Dinosaurs abundant; first birds; gymnosperms (cycads and conifers) dominant; Pangea splits; tropical conditions
	Triassic (248–206)		Archosaurs dominate therapsids after extinction; reptiles in sea and air; true mammals and modern corals appear
Paleozoic (543–248)			
	Permian (290–248)		Pangea forms; swamps dry; reptiles diversify; amphibians decline; mammal-like therapsids evolve
	Carboniferous (354–290)		Extensive coal swamps; large conifers; reptiles evolve from amphibians; insects diversify; oxygen levels very high
	Devonian (417–354)		Diversification of fishes and sharks; ammonoids appear; first seed plants (gymnosperms); first amphibians on land
	Silurian (443–417)		Primitive plants invade the land; first vascular plants (*Cooksonia*); early arthropods on land; armored fishes appear
	Ordovician (490–443)		Rugose and tabulate corals and stromatoporid reefs are home to abundant arthropods; first vertebrates
	Cambrian (543–490)		Explosion of marine life; rapid diversity; most phyla now in place; marine invertebrates and algae dominate seas
Precambrian (4600–543)			
	Proterozoic Eon (2500–543)		Rapid build-up of atmospheric oxygen; banded iron formations; first protist life; barren supercontinent Rodia
	Archaen Eon (3800–2500)		Geological record begins; primitive cyanobacteria appear and increase in number; stromatolite reefs form
	Hadean Eon (4600–3800)		The "rockless" eon; the Solar System coalesces into planets; a thin crust begins to form over the molten Earth

BIODIVERSITY GRAPH

Earth's biodiversity, or the number of families that existed throughout geological time, is shown here in graphical form. Following the dramatic explosion of life in the Cambrian, it is interesting to note that the graph does not show a slow, steady increase in the number of families over time, as one might expect. Its shape is highly dependent on geological and tectonic happenings on Earth and extraterrestrial events.

Sharp, sudden drops in the total number of families represent mass extinction events of mostly uncertain causes, but possibly due to extensive glaciation episodes, meteorite impacts or large-scale volcanic events. Events of this magnitude obviously affected the climate sufficiently to break the normal operation of the food chain, with dramatic consequences for certain families. The major extinctions are marked on the graph. A significant drop in the number of families on Earth during the Triassic period appears to be due to a combination of the calamitous mass extinction at the end of the Permian (the largest on record) and the coming together of the continents to form the Pangean supercontinent, with a consequent biodiversity loss due to interspecies competition.

Since the Jurassic, family numbers have been rising steadily, accompanying the break-up of the Pangean supercontinent into the present-day continental arrangement. The last major setback in this increasing trend occurred at the end of the Cretaceous period (the KT boundary), possibly due to a large meteorite strike in the Yucatan Peninsula, Mexico. This mass extinction saw the end of the dinosaurs and opened the way for the mammals to rise to a position of dominance—and from them arose the human race.

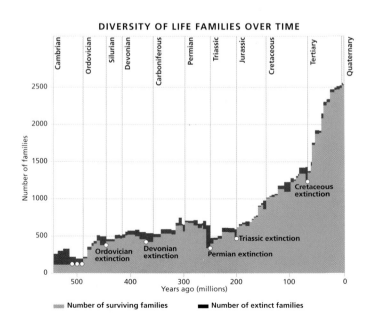

DIVERSITY OF LIFE FAMILIES OVER TIME

Cambrian · Ordovician · Silurian · Devonian · Carboniferous · Permian · Triassic · Jurassic · Cretaceous · Tertiary · Quaternary

Number of families

Cretaceous extinction

Triassic extinction

Ordovician extinction

Devonian extinction

Permian extinction

Years ago (millions)

▨ Number of surviving families   ▮ Number of extinct families

SOME FOSSIL SITES

EDIACARA

Location: Ediacara Hills, South Australia
Period: Proterozoic eon, 570–543 million years old
Finds: Abundant soft-bodied intertidal invertebrate marine lifeforms, known collectively as Ediacaran Fauna; they include jellyfishes, segmented worms such as *Dickinsonia*, featherlike soft corals or sea pens, early arthropod-like forms such as *Parvancorina*, and other strange forms, unrelated to present-day life, not easily classified.

MISTAKEN POINT

Location: Avalon Peninsula, Newfoundland, Canada
Period: Proterozoic eon, 565 million years old
Finds: An important assemblage of strange jellyfish-like and other invertebrate deep-water marine lifeforms, buried and preserved by an eruption of volcanic ash. Most do not resemble modern-day lifeforms, with some scientists even assigning them to a new kingdom of life. The most common fossils are the bizarre spindle-shaped forms.

BURGESS SHALE

Location: Yoho National Park, British Columbia, Canada
Period: Cambrian, 505 million years old
Finds: Excellently preserved, delicate, soft-bodied shallow marine invertebrates swept off the edge of the continental shelf and rapidly buried. Over 140 species have been found, including arthropod-like armored and spiked forms. One of these is the 2-foot (60-cm) long carnivorous *Anomalocaris*, with specialized grasping appendages.

HUNSRUCHSCHEIFER FAUNA

Location: Bundenbach, Wissenbach, Gemunden, Germany
Period: Devonian, 370 million years old
Finds: Hard and soft parts of organisms preserved as pyrite within shale, revealing exquisite detail when X-rayed. Over half the known specimens are of the arthropod *Mimetaster*, a deposit-feeder with eyes on stalks and two pairs of strong walking legs. Not found elsewhere, it may be related to the Burgess Shale's *Marrella*. Other fossils include sea stars, trilobites, other arthropods and cephalopods.

MAZON CREEK

Location: Illinois, USA
Period: Carboniferous, 300 million years old
Finds: A river delta advancing over extensive swamps preserved over 400 plant species and 320 animal species. Flora include horsetails, ferns and club mosses. Fauna include jellyfishes, mollusks, arthropods, fishes and amphibians. A strange arthropod, *Tullimonstrum*, with its spadelike tail, is Illinois' state fossil. Mazon Creek is also a Lagerstatten, a deposit where soft body parts are preserved as carbon films.

KAROO BEDS

Location: Karoo Basin, South Africa
Period: Permian to Triassic, 280–210 million years old
Finds: Karoo witnessed the evolution of mammals from reptilian ancestors. It was a complex of rivers draining into a sea that preserved fossils over a continuous period of more than 50 million years. This allowed evolutionary history of many groups, such as dicynodonts, to be tracked. Toward the end of the Triassic, early dinosaurs appeared.

HOLZMADEN AND SOLNHOFEN

Location: Stuttgart and Bavaria, Germany
Period: Jurassic, 150 and 190 million years old
Finds: Holzmaden has abundant marine reptiles, complete skeletons of ichthyosaurs and pliosaurs. Some are so perfectly preserved they contain stomach contents and unborn young. Skin and tissue are also often preserved. The most important fossils from Solnhofen limestone are the amazing dinosaur–bird *Archaeopteryx* specimens.

DINOSAUR NATIONAL MONUMENT

Location: near Vernal in Utah and Colorado, USA
Period: Jurassic, 150 million years old
Finds: Remains of 100 dinosaurs representing 10 species that were washed into the sandbar of a large ancient river. They include the plant-eating *Diplodocus*, *Apatosaurus*, *Camarasaurus*, *Barosaurus*, *Stegosaurus*, *Dryosaurus* and *Camptosaurus*, and carnivorous *Allosaurus* and *Ceratosaurus*. Bones can be seen embedded in the quarry face.

AUSTRALIAN OPAL FIELDS

Location: Great Australian Basin, Australia
Period: Early Cretaceous, 110 million years old
Finds: Fossils from a shallow sea covering much of Australia have been replaced by opaline silica and often precious opal. Most abundant are bivalves and gastropods; also found are belemnites (straight-shelled nautiloids), bones and even whole skeletons. Most bones are from plesiosaurs, pliosaurs and ichthyosaurs. Rare land dinosaur and mammal fossils have been found at Lightning Ridge and Andamooka.

EGG MOUNTAIN

Location: Montana, USA
Period: Late Cretaceous, 80–70 million years old
Finds: The Egg Mountain site revealed that baby dinosaurs were hatched in rookeries and stayed in the nest for extended periods while parents brought back food. Rookeries of up to 40 bowl-shaped depressions, with up to 25 grapefruit-sized eggs, belong to *Maiasaura* (Good mother lizard), a large plant-eating duck-billed dinosaur. Herd behavior protected them from predators such as *Troodon*.

DINOSAUR PROVINCIAL PARK

Location: Drumheller, Alberta, Canada
Period: Late Cretaceous, 75 million years old
Finds: Specimens are visible eroding from rock and on display in the Royal Tyrrell Museum. They are duck-billed and horned dinosaurs, including bone beds of thousands of *Centrosaurus* individuals that migrated across the continent in search of food. Those that lost their lives during dangerous river crossings were swept downstream into bone beds.

GRUBE MESSEL

Location: Darmstadt, Germany
Period: Eocene, 49 million years old
Finds: Lake floor shales mined for oil were found to contain fish, crocodile, snake, bat, horse, tapir, turtle and various plant fossils. Preservation is excellent, with delicate detail of skin, fur, feathers, internal organs and gut contents. Beetles have even been found with the iridescent color of their wing covers still perfectly preserved.

LAKE TURKANA

Location: Lake Turkana Basin, Kenya
Period: Miocene to Holocene, 25 million years–present
Finds: Over the past 25 million years, rivers have washed fossils and sediments into the Lake Turkana Basin from the surrounding highlands. Interspersed volcanic ash beds allow dating of the thick sequence. The skeleton of an 11-year-old boy (*H. ergaster*) was dated at 1.6 million years old. Stone tools, 2.3 million years old, were also found here.

OLDUVAI GORGE

Location: Eastern Serengeti Plains, Tanzania
Period: Pliocene, Pleistocene, 2.1-0.015 million years old
Finds: Over 50 hominid remains, including *Homo habilis*, *Australopithecus* and *H. erectus*, together with their stone tools and campsites, have been found at Olduvai Gorge, buried and preserved by ash from nearby volcanoes. This site was discovered by Louis Leakey in the 1930s and affords an insight into stone tool and hominid evolution through a series of progressively younger beds.

RANCHO LA BREA TAR PITS

Location: Los Angeles, USA
Period: Pleistocene, Holocene, 40,000 years to present
Finds: Famed for its beautifully preserved large mammals, including bison, horses, saber-tooth cats, wolves, coyotes, sloths, camels, antelopes and bears. Predator fossils outnumber larger prey fossils by six to one, suggesting that luckless animals trapped in the sticky tar pools became the target of predators anticipating an easy meal. The predators themselves were trapped and, in turn, became the victims.

Glossary

A'a A type of basaltic lava that has a jagged broken surface.

Abrade Wearing away or rounding during transportation.

Active volcano A volcano producing regular eruptions of gas and lava. Eruptions may be weeks to centuries apart.

Adularescence A type of sheen in minerals; the soft-glowing, cloudy effect seen in moonstone, a type of feldspar.

Aftershock A tremor that follows a large earthquake and originates at or near the hypocenter of the initial quake.

Agate An ornamental cryptocrystalline quartz mineral, a variety of fine-grained chalcedony. It occurs in many colors and is often striped or concentrically patterned.

Aggregate A mass of interlocking mineral crystals of varied size, often the result of crystals forming in a confined space.

Alluvial fan A fan-shaped deposit of sediment formed at an abrupt decrease in stream gradient, such as at the foot of a mountain.

Anthracite The most compact form of coal, with a higher carbon and lower volatile content than bituminous coal.

Anticline An arch-shaped fold in layers of sedimentary rock.

Ash Fine pieces of rock and lava ejected explosively during volcanic eruptions. Small particles may travel great distances.

Asthenosphere A layer in Earth's upper mantle that is semi-soft like modeling clay. It lies below the hard, rigid crust.

Atom The smallest unit that can be called a chemical element. All things on Earth are made up of atoms.

Bacteria Single-celled, microscopic lifeforms with no true nucleus. They belong to the kingdom Monera.

Baltica One of the ancient continental fragments that broke away from from Rodinia, consisting of what are now the landmasses around the Baltic Sea.

Bedrock The solid rock found below soil and weathered surface rock fragments.

Black smoker A vent situated on an ocean ridge that emits hot, mineral-laden water, and is home to unusual lifeforms.

Butte A flat-topped, steep-sided hill often found in desert areas. Its width is narrower than its height.

Calcareous Containing calcium carbonate. For example, some shells and corals, and rocks such as limestone and marble.

Caldera A large, circular depression formed when the top of a volcano collapses into its empty magma chamber.

Canyon A deep, steep-sided valley formed by river erosion.

Cenozoic era A geological era from 65 million years ago to the present, comprising the Tertiary and Quaternary periods.

Cephalopod A class of marine mollusks, characterized by tentacles. Forms include those with coiled or straight shells (ammonites and nautiloids), as well as octopus and squid.

Chitin A strong, rigid, lightweight material that is a component of the exoskeleton of many arthropods.

Clay A fine-grained sediment formed by the chemical breakdown of rocks. It is moldable when wet and hard when dry. It can be baked to make china, pottery, tiles and bricks.

Cleavage The tendency of certain minerals to break along set planes of weakness related to their crystal structure.

Cnidaria A phylum of animals including jellyfish, anemones and corals. They are characterized by stinging tentacles.

Coal A sedimentary rock formed by the compression of plant remains. One of the fossil fuels, it can be burned for energy.

Compound A chemical substance made up of more than one element. Most minerals are compounds.

Concretion A usually rounded, hard, mineralized mass, often concentrically banded and typically with a fossil at its core.

Contact metamorphism The transformation of one type of rock into another, mostly as a result of heating.

Continent One of Earth's seven main landmasses: Africa, Antarctica, Asia, Australia, Europe, North America and South America. These landmasses have edges beneath the ocean known as continental shelves.

Continental shelf The shallow, submerged, fringing part of a continent, to 700 feet (200 m) deep. Results initially from subsiding rift valley sides, but is later modified by events on collision margins, which add ocean-floor sediment and volcanic islands to the shelf edge. Coral reefs build on tropical shelves.

Convection currents (mantle) Movement within the mantle caused by heat transfer from Earth's core. Hot rock rises and cooler rock sinks. This movement is most likely responsible for the motion of Earth's tectonic plates.

Convergent margin A boundary between two tectonic plates that are moving toward each other.

Coprolites Fossilized animal droppings. These are trace fossils that reveal much about diet of the creatures they belong to.

Corallite The small cup of calcium carbonate in the coral skeleton in which the coral animal, or polyp, lives.

Core Earth's dense center. It consists of a solid inner core and a molten outer core, both of which are made of iron.

Core sample A long column of rock that has been extracted from the ground by drilling with a hollow drill. Geologists use core samples to study rocks, ice or soil beneath the surface.

Crater A circular depression formed as a result of a volcanic eruption (volcanic crater) or by the impact of a meteorite (impact crater).

Crater lake A water-filled crater. It may be filled on a seasonal or permanent basis.

Cretaceous period The last period of the Mesozoic era, from 144 to 65 million years ago. Dinosaurs died out by its end.

Crust The outer layer of Earth. There are two types of crust: continental crust, which forms the major landmasses; and oceanic crust, which is thinner and forms the seafloor.

Crystal A solid mineral form with a characteristic internal molecular structure enclosed by faces that meet at definite and specific angles related to the structure of the substance.

Dendritic habit Having a branched, fernlike appearance.

Diatom A microscopic, single-celled protist growing in marine and fresh water; it has hard cell walls composed of silica, the result of the organism extracting dissolved silica from water.

Dike A sheet of igneous rock formed when magma is injected into a vertical or dipping crack.

Dinoflagellate Single-celled organism (protist) with a cellulose shell. They are important marker fossils.

Divergent margin A boundary between two tectonic plates that are moving apart on either side of a spreading ridge.

DNA Deoxyribonucleic acid, found in living cells, is the carrier of information that tells an individual organism how to build itself. Closely related organisms have a greater similarity in their DNA than those that are more distant, so DNA analysis can provide information on the relationships between species.

Dormant volcano A volcano that is not currently active but that could erupt again at any moment.

Earthquake A sudden, violent, release of energy in Earth's crust that generally occurs at the edges of tectonic plates.

Ecosystem An interdependent community of plants, animals and other organisms, and the environment in which they live; for example, wetland, desert, pond or coral reef.

Element A substance that contains only one kind of atom. For example, gold, sulfur and diamond are elemental minerals.

Eocene epoch A geological epoch (within the Tertiary period) between 54.8 million and 33.7 million years ago.

Epicenter The point on Earth's surface that is directly above the hypocenter, or starting point, of an earthquake.

Era A division of time in Earth's history. Geologists divide eras into periods, then, in turn, into epochs and then ages.

Erosion The gradual wearing away of rock or landscape by water, ice or wind.

Eruption The volcanic release of lava and gas from Earth's interior onto the surface and into the atmosphere.

Euramerica The late Silurian–Devonian continent formed by the collision of the Baltica craton and Laurentia and the closure of the Lapetus Ocean. It is also known as Laurussia and the Old Red Continent because of its distinct red sandstone deposits.

Evaporite A non-metallic sedimentary mineral precipitate formed by the evaporation of saline water in desert lakes.

Extinct volcano A volcano that has shown no sign of activity for a long period and is considered unlikely to erupt again.

Fault A fracture in Earth's crust along which displacement has occurred. Faults may be normal, reverse or strike-slip.

Fault margin A crack in rock layers created by the rocks shifting in opposite directions or at different speeds.

Fissure A fracture or crack in the ground. In volcanic areas, eruptions may occur as a line of vents along a fissure.

Flood basalt A flow of basaltic lava that spreads over a large area. It fills in the valleys and then forms a basalt plateau.

Flowbanding A patterned effect that results from crystals in some magmas or lavas responding to flow movement by aligning themselves in bands; for example, rhyolite.

Flowstones Common in limestone caves and thermally active regions. They precipitate from mineral-saturated water forced to the surface where it cools and deposits the minerals.

Fluorescent mineral A mineral that glows under ultraviolet light in a color often different from its natural color.

Fold A bend in rock layers caused by crustal movement.

Foliation The arrangement of minerals in crude planes, caused by regional metamorphism, which produces rock with a striped appearance and bands of alternating composition.

Foraminifera (forams) Single-celled marine protists with calcium carbonate shells, which may build marine deposits.

Fossil Any preserved evidence of pre-existing life. It may be the remains of a plant or animal that has turned to stone or has left its impression in rock.

Fossil fuel A fuel that formed when plant remains were compressed under sedimentary rock layers. The most common fossil fuels are coal, oil and natural gas.

Gastroliths Stones that some dinosaurs and birds swallowed to aid digestion by grinding up food in the stomach.

Gemstone Any rock or mineral used for personal adornment; usually uncommon, transparent, colorful or brilliant material that can be cut and polished.

Genus (plural genera) A group of closely related species.

Geode A rock cavity, lined with concentric mineral bands and/or crystal formations pointing into the center.

Geological period A specific, dated division of geological time. They are further broken down into epochs and ages.

Geological time The vast length of time that stretches from the formation of Earth to the present day; divided into eons, then eras, then periods, epochs and ages.

Geologist A person who studies geology or is involved in the exploration of Earth for economic rocks and minerals.

Geology The study of Earth. Rocks, minerals and fossils give some of the clues to Earth's history.

Geothermal energy Energy that can be extracted from Earth's interior heat, whether from hot rocks, hot water or steam.

Geyser A surface vent that periodically spouts a fountain of boiling water; for example, Old Faithful geyser, USA.

Glaciation The effect of a moving ice mass on a landscape, resulting in erosion, the gouging of U-shaped valleys and fjords, and the deposition of ridges and sheets of rock debris.

Glacier A large mass of ice formed by the buildup of snow on a mountain or a continent. The ice moves slowly downhill.

Gondwana The southern supercontinent fragment comprising New Zealand, Antarctica, Australia, South America, Africa and India. It existed as a separate landmass from 650 million years ago and only began to break up 130 million years ago. It influenced climate and biodiversity.

Habit The exterior shape of a single crystal or a group of crystals of the same mineral.

Hominid A member of the family Hominidae, which includes both extinct and modern forms of humans.

Hoodoo A tall column of rock formed by erosion.

Hot spot A persistent zone of melting within Earth's mantle.

Hot spot volcano A shield volcano that forms on a plate as it passes over a stationary hot spot. Plate movement causes a line of such volcanoes to form, becoming progressively older.

Hypocenter The place within Earth where energy in strained rocks is suddenly released as earthquake waves.

Ichthyosaurs Marine reptiles that adapted to life in the sea by becoming dolphinlike in appearance. They lived at the same time as dinosaurs and gave birth to live young at sea.

Igneous rock Rock that forms when magma cools and hardens. Intrusive igneous rock solidifies underground and extrusive igneous rock solidifies above ground.

Impermeable rock Rock through which liquids cannot pass.

Index fossil A fossil species that is restricted to a geological layer of a particular age.

Intrusion A large mass of rock that forms underground when magma is injected into other rocks and then slowly hardens.

Invertebrate General term for animals without backbones, such as worms, mollusks and insects.

Iridescence A type of sheen; a rainbow effect, as seen in mother-of-pearl and labradorite feldspar.

Island arc An arc-shaped chain of volcanic islands that forms above a subducting seafloor.

Jurassic period The middle period of the Mesozoic era, from 206 to 144 million years ago. Many new dinosaurs appeared in this period.

Kimberlite A type of olivine-rich, ultrabasic rock erupted from the mantle. Most diamonds come from kimberlites.

Laccolith A mushroom-shaped body of volcanic rock formed when rising magma pushes rock layers upward.

Lahar A mudflow created by a volcanic eruption.

Laurasia One of the two continents that formed when the supercontinent Pangea separated. It includes Europe, North America and Asia (not India). Similarity of plants and animals of these countries is explained by this former connection.

Laurentia One of the ancient continental fragments of the Rodinia supercontinent. It includes much of North America.

Lava Molten rock that has erupted onto Earth's surface.

Lava bomb A large lump of lava, usually more than 1.25 inches (32 mm) across, that is thrown out of a volcano.

Lava dome A mound of thick, sticky lava that grows at the top of, or on the flanks of, a stratovolcano.

Lava flow A river of lava that erupts from a volcano and runs over surrounding land, usually following river valleys.

Lava tube An underground tube that forms when the surface of a lava flow solidifies while the inside remains molten.

Liquefaction The change of sediment or soil into a fluid mass as a result of an earthquake.

Lithosphere The rigid outer part of Earth, consisting of the crust and the uppermost part of the mantle.

Lophophorates Marine invertebrates that strain micro-organisms from the water; they include the brachiopods.

Luster The way light falling on the surface of a mineral is absorbed or reflected (compare sheen).

Magma Hot, liquid rock or a mush of liquid rock and crystals found beneath the surface of Earth. When magma erupts onto Earth's surface, it is called lava.

Magma chamber A pool of magma in the upper part of the lithosphere from which volcanic materials may erupt.

Magnitude The strength of an earthquake, based on the amount of energy released. Seismologists measure magnitude using the Richter scale.

Mantle The layer between Earth's crust and outer core; it is semi-solid rock with the consistency of modeling clay.

Massive A manner of occurrence of some minerals in which many mineral grains are intergrown to form a solid mass.

Matrix The rock in which a fossil or crystal is embedded.

Melt A quantity of molten rock such as magma or lava.

Mesa A wide, flat-topped hill with steep sides. The width of a mesa is greater than its height (see also butte).

Mesozoic era The geological era between 248 million and 65 million years ago, comprising the Triassic, Jurassic and Cretaceous periods. Known as the "age of dinosaurs," it was also the era when the first birds and mammals appeared, flowering plants became dominant, and Pangea broke apart.

Metal Any of a number of elements that are shiny, moldable and will conduct electricity. Many metals are found in minerals as compounds.

Metamorphic rock Rock formed by the transformation of a pre-existing rock as a result of heat and/or pressure.

Metazoans Multicelled animals, comprising all animals apart from the sponges and the single-celled protozoans (the latter often classified with algae).

Meteor A streak of light in the night sky caused by a lump of rock entering Earth's atmosphere from space. Before the rock enters the atmosphere it is known as a meteoroid. If it lands on Earth's surface, it is called a meteorite.

Meteorite A piece of planetary material from outside Earth that has fallen through the atmosphere onto Earth's surface.

Microfossil Tiny fossils such as diatoms and radiolaria. Often widespread and numerous, they can make good index fossils.

Mid-oceanic ridge A long, submerged mountain range that runs through the ocean between continents; the boundary where two plates are pulled apart and new plate material is added. Formed by the upwelling of hot basaltic magma.

Mineral A naturally formed solid with an ordered arrangement of atoms; found in Earth's crust.

Miocene epoch A geological epoch (occurring within the Tertiary period) between 23 million and 5 million years ago, during which time many animals of modern form, including the modern apes, evolved.

Molecule A cluster of atoms formed when one or more types of atoms bond.

Mudflow A river of ash, mud and water set off by a volcanic eruption or earthquake. Those triggered by volcanoes are also known as lahars.

Native element An element that exists alone and not in combination with another element. Sulfur, diamond, platinum and gold are examples of elemental minerals.

Native metal A metal that occurs in its elemental form, such as gold, silver, or copper (unlike aluminum, iron or tin, which are found only as compound ores).

Natural selection An evolutionary theory proposing that only those members of a population best suited to their environment survive and reproduce, so that over a long period of time certain genes become more common in later generations, eventually leading to evolution of new species.

Normal fault A fracture in rock layers, where the upper side has moved downward relative to the other side along a plane inclined between 45 and 90 degrees.

Ocean trench A deep, narrow undersea valley formed when the oceanic crust of one tectonic plate collides with and dives beneath the crust of another plate.

Ore A mineral or rock that contains a particular metal in a concentration that is high enough to make its extraction commercially viable. Hematite and iron ore are examples.

Organic Being derived from living organisms. Amber, pearls, shell, coral, coal and some cherts and limestones are described as being of organic origin.

Ornamental stone An attractive gemstone that is not considered precious but can be used for jewelry or other types of ornamental use, such as carving and turning.

Outcrop The part of a rock formation that is exposed at Earth's surface and not covered by sand or soil.

Pahoehoe A type of lava with a smooth, ropelike surface.

Paleocene epoch A geological epoch (within the Tertiary period) between 65 million and 54.8 million years ago.

Paleontology The study of plant and animal fossils.

Paleozoic era The geological era between 543 million and 248 million years ago, comprising the Cambrian, Ordovician, Silurian, Devonian, Carboniferous and Permian periods. During this era, land plants developed (many forming today's coal deposits), marine invertebrates and fishes were dominant, amphibians became the first air-breathing vertebrates on land, and the supercontinent of Pangea was formed.

Pangea A supercontinent; a single landmass that formed as a result of the collision of all the continents during the Permian period and broke up during the Jurassic into Laurasia and Gondwana. These, in turn, broke into the present continents.

Pannotia Sometimes called the Vendian supercontinent that evolved from Rodinia. It assembled during the Cambrian while breaking apart in other regions. The three main daughter continents were Gondwana, Laurentia and Baltica.

Pegmatite Element-rich fluid granite that penetrates cracks in the rock in and around granite intrusions, there producing large and beautiful crystals of minerals, such as quartz, feldspar, topaz, tourmaline and beryl.

Period A geological time unit, a subdivision of an era. The periods are Precambrian, Cambrian, Ordovician, Silurian, Devonian, Carboniferous, Permian, Triassic, Jurassic, Cretaceous, Tertiary and Quaternary.

Petrification The cell-by-cell replacement of decaying organic matter such as bone or wood by water-borne minerals from surrounding solutions, resulting in rock-hard mineral casts of plants and animals.

Phosphatic Rare rocks that contain abundant phosphate minerals. They may be bone beds or guano (bird droppings).

Phosphorescent mineral A mineral that continues to glow for a short time in the dark after the ultraviolet light is turned off.

Photosynthesis The process by which green plants use the energy of sunlight to convert water and carbon dioxide to sugar and starch, while releasing oxygen into the air.

Phylum A major division of the animal and plant kingdoms consisting of organisms constructed on a similar general plan and thought to be descended from a common ancestral form.

Pillow lava Lava that forms rounded mounds by cooling quickly after erupting under water or flowing into water.

Plate movement The movement of Earth's tectonic plates, probably caused by convection currents in the mantle.

Plate tectonics The process of movement of crustal plates on Earth's surface by which continents and oceans are created.

Play-of-color An optical effect seen in opal that is caused by light refracting and which creates vivid patches of color.

Pleistocene epoch A geological epoch between 1.8 million and 10,000 years ago during which time ice sheets advanced across northern Europe and North America, giant mammals roamed the forests and plains, and modern humans appeared.

Plesiosaurs Large, fish-eating, mainly long-necked marine reptiles that lived during the Mesozoic. They had four paddlelike flippers. This group included *Plesiosaurus* and *Elasmosaurus*.

Pliosaurs Marine reptiles that first appeared in the early Jurassic. They had large heads, short necks, thick, powerful bodies and very strong teeth and jaws. They were the most formidable marine predators of the Mesozoic. *Kronosaurus* was a pliosaur.

Plug A column of volcanic rock formed when lava solidifies inside the vent of a volcano. Ship Rock, USA, is an example.

Plume A rising column of hot rock in the mantle within which melting can take place. The term can also apply to a large column of ash above a volcano.

Pluton A mass of igneous rock that rose buoyantly from depth as a melt to intrude and cool near surface rocks as a balloon-like body.

Polyp A tiny colonial anemone-like animal that removes calcium carbonate from the sea to build coral reef structures. It is characterized by stinging cells in its tentacles.

Primary rock Rock formed when magma cools. Basalt crystallizing at the mid-oceanic ridge is an example.

Primary wave A seismic wave, also known as a P-wave, which compresses and expands rocks as it travels through them. It is called a primary wave because it is the wave that arrives first during an earthquake, before the secondary wave.

Pterosaurs Flying reptiles, such as *Ramphorhynchus* and the giant *Quetzalcoatlus*, which evolved during the late Triassic. They were distant relatives of the dinosaurs.

Pumice A light-colored, glassy volcanic rock that contains many cavities. It is so light that it can float in water.

Pyroclastic flow A dense, heated mixture of volcanic gas, ash and rock fragments that travels at great speed down volcanic slopes. It forms as a result of the collapse of an eruption column or a lava dome.

Radiolaria A single-celled planktonic animal-like protist with a silica skeleton. Their remains may accumulate to form chert.

Regional metamorphism Large-scale transformation of one rock type into another as a result of heat and pressure due to plate collisions and the eventual formation of large ranges of mountains.

Reverse fault A fracture in rock layers, where the top side has moved upward relative to the other side along a plane inclined between 45 and 90 degrees.

Rift valley A wide valley that forms as Earth's crust is stretched apart and a central section drops downward as a result of normal faulting.

Rifting The process of splitting continental plates. Upwelling magma pulls the continent apart; as the rift widens, water floods in to form a new ocean.

Rock A solid mass usually made up of minerals and/or rock fragments. Classed as igneous, metamorphic and sedimentary.

Rodinia The Precambrian supercontinent and the earliest that science can piece together—but probably not Earth's first. It fragmented toward the end of the Precambrian into north China, Siberia and Pannotia.

Rugose coral A group of solitary and colonial corals that became extinct at the end of the Permian. They were replaced by the modern scleractinian corals.

Scleractinian coral Colonial corals of a modern form that arose in the Triassic and replaced the rugose and tabulate corals.

Secondary wave A seismic wave, also known as an S-wave, that moves rocks from side to side as it passes through them. It is called a secondary wave because it is the second type of wave to arrive during an earthquake.

Sediment Weathered pieces of rocks or plant or animal remains that are deposited at the bottom of rivers and lakes by water, wind or ice.

Sedimentary rock Rock formed near Earth's surface from pieces of other rocks or plant or animal remains, or by the buildup of chemical solids.

Seismic Related to an earthquake or tremor.

Seismic waves Energy waves that travel through Earth after an earthquake, often causing great destruction.

Seismogram A graph or computer image that depicts Earth tremors as wavy lines. It is produced by a seismograph.

Seismograph An instrument that detects, magnifies and records Earth's vibrations.

Seismology The study of Earth tremors, whether natural or artificially produced.

SEM (Scanning Electron Micrograph) A very high-powered microscope that uses electron beams to record detailed images of the subject being studied.

Septa Thin plates of calcium carbonate that radiate inward from the wall of the coral cup in which the polyp lives. The septa support the concertina folds of the polyp's stomach.

Sheen The reflection of light from structures, such as lamellae, or cracks, within a mineral (compare luster).

Shield volcano A wide, low volcano formed by continuous fluid basaltic lava flows. This type of volcano looks like a shield when viewed from above.

Siliceous Containing silica; any rock that is silica or quartz-rich is termed siliceous or acidic. For example, granite and chert.

Sill A sub-horizontal band of igneous rock formed when magma intrudes and solidifies between parallel rock layers.

Solution A mixture of two or more chemical substances. It may be a liquid, solid or gas.

Species A group of similar plants or animals that can reproduce with other members of the group, but not reproduce with members of other groups (compare genus).

Specific gravity The relative weight of a mineral compared to the weight of an equal volume of water; used as a means of mineral identification.

Stratigraphic column A diagram that shows the chronological sequence of rock units.

Stratovolcano A steep-sided, cone-shaped volcano that occurs at a subducting plate boundary where the downgoing plate melts, allowing andesitic magma to rise and pierce the overriding plate. These volcanoes are characterized by violent eruptions and explosions.

Stratum (plural strata) A stratum is the same as a bed. Rocks are stratified if they display bedding planes.

Streak The mark of powdered mineral left when a mineral is rubbed against unglazed white porcelain; the color of the streak is used in mineral identification.

Streak test A test that involves rubbing a mineral across an unglazed porcelain tile to produce a powder. The color of the powder left by the mineral can help identify it.

Striations Scratches in bedrock from the passing of a glacier, or parallel lines on the faces of some crystals.

Strike-slip fault A fault along which rocks have moved sideways. It is sometimes called a transform fault.

Subduction The sliding of a dense oceanic plate under the edge of a more buoyant plate, which is usually continental. Eventually, the sinking plate will be remelted into magma.

Surface wave A seismic wave that travels along Earth's surface. It arrives after primary and secondary waves and moves up and down or from side to side.

Symbiosis A relationship between two different species that may be beneficial to one or both parties. For example, the relationship between insects and flowering plants.

Syncline A basin-shaped fold in layers of sedimentary rock.

Tabulate coral A primitive colonial coral that became extinct along with the rugose corals in the Permian mass extinction.

Tectonic plates Rigid pieces of Earth's lithosphere that move over the semi-solid asthenosphere.

Tectonic uplift The raising up of rock as a result of plate movements, such as is occurring in the Himalayas today.

Tethys Sea The body of water partially enclosed by the C-shaped Pangean supercontinent. It was closed when Pangea split into Laurasia and Gondwana.

Theropods A group including all the meat-eating dinosaurs, from small predators such as *Coelurus* and *Compsognathus* to huge hunters such as *Tyrannosaurus* and *Spinosaurus*. These saurichsians were all bipedal.

Thrust fault A fracture in rock where the upper side rides over the lower side at an angle of less than 45 degrees.

Trace fossils Fossilized signs and remains of an animal's activity, such as footprints, bite marks, burrows, nests, eggs, gastroliths (stomach stones) and droppings.

Transform fault A fault or plate margin along which rocks move in opposite directions or at different speeds.

Triassic period The first period of the Mesozoic era, from 248 to 206 million years ago, when Earth's landmass comprised only one giant supercontinent, Pangea. Dinosaurs first appeared in this period.

Tsunami A Japanese word for a sea-wave produced by an earthquake, landslide or volcanic blast. It reaches its greatest height in shallow waters before crashing onto land.

Vent A pipe inside a volcano through which lava and gas move from the magma chamber and erupt on the surface.

Volcanic plug A stump of hard igneous rock that remains after a volcano has been worn away by erosion.

Volcano A typically circular, cone-shaped landform built from the accumulation of lava flows and ash-falls.

Volcanologist A scientist who studies volcanoes, active as well as inactive. Volcanology is a branch of geology.

Weathering The disintegration of rocks and minerals as a result of the freezing and thawing of ice, the action of chemicals in rainwater or the growth of plant roots.

Zooxanthellae Unicellular yellow-brown protists (dinoflagellates) that live within the polyp of the reef-building coral in a symbiotic relationship.

Index

Credits

PHOTOGRAPHS

t=top; l=left; r=right; tl=top left; lc=top center; tr=top right;
cl=center left; c=center; cr=center right; b=bottom; bl=
bottom left; bc=bottom center; br=bottom right
ACD = Artville; AMNH = American Museum of Natural History;
APL/CBT = Australian Picture Library/Corbis ; AUS = Auscape
International; COR = Corel Corp.; DS = Digital Stock; GI = Getty Images;
GSL = Gem Studies Laboratory; N_EO = NASA/Earth Observatory; N_J =
NASA/JPL; N_V = NASA/Visible Earth; NGS = National Geographic Society;
NHM = Natural History Museum, London; PD = Photodisc; PE =
PhotoEssentials; PL = photolibrary.com; SP=Seapics.com; TPL/SPL =
photolibrary.com/ Science Photo Library; TSA = Tom Stack & Associates;

1c GSL/Bill Sechos 2c PL 4c DS 6–7c APL/Corbis 8–9c Jim Frazier 10c GI cl,
cr APL/Corbis 11c, cr PL cl PD 12c APL/Corbis 14–15c APL/Corbis 16c,
cl APL/Corbis 17c, cl APL/Corbis 18b GI tr APL/Corbis 19cr APL/Corbis 20bl,
cr PL 21br Jesse Fisher tr APL/Corbis 22bl PL br Jesse Fisher 24tr PL 25t
APL/Corbis 26tr Brian M England 27tr APL/Corbis 28b APL/Corbis 29br,
tr APL/Corbis 30bc PL bl N_V tr APL/Corbis 32tr APL/ Corbis 33tc APL/Corbis
34bl APL/Corbis tr GI 35cr, tr APL/Corbis 36bl PL br TSA 37cr TSA 38bl
APL/Corbis cr PL 39t APL/Corbis 40bl APL/Corbis br PL 41b APL/Corbis tl GI
tr PL 42b APL/Corbis 43cr Artville tc, tl PL 44b GI 45bl GI 46bl, c GI 47bl GI
tl PL 48b APL/Corbis 49cl, c APL/Corbis 50bl, br PL tr APL/Corbis 51br
APL/Corbis br PL 52b APL/Corbis 53cr GI tl PL tr APL/ Corbis 54tr APL/Corbis
55b APL/Corbis 56br COR 57b NHM t GI 58b, tr APL/Corbis 59bl PL br
APL/Corbis 60–61c GI 62c, cr GI cl APL/Corbis 63cl AUS cr PL 64bl, tr
APL/Corbis 66br AUS/ Reg Morrison tr AUS/Lee McElfresh 67br APL/Corbis
68bl APL/Corbis tr NGS/O Louis Mazzatenta 69c GI 72r AUS 73br AUS/Reg
Morrison t PL 76bl, tr PL 77cr APL/Corbis 78bl AUS/Reg Morrison r
SP/Marty Snyderman 79tr AUS/ Jean-Marc La Roque 82bc, cl APL/Corbis cr
GI 84bl, cr APL/ Corbis 88bl APL/Corbis r AUS/Reg Morrison 89tr DS 92bl,
tr APL/Corbis 93t PL 94b GI 95br, tl PL 98bl, tr APL/Corbis 99br APL/Corbis
tr GI 100bl TSA cr GI 101br GI tr PL 104–105c APL/Corbis 106c, cl APL/Corbis
cr PL 107cl APL/Corbis cr PL 108bl, tr APL/ Corbis br GI 110b APL/Corbis tr
PL 111bl, t PL br APL/Corbis 112b, t PL 113b AUS t APL/Corbis 115br, tl
APL/Corbis tr PL 117c PL 118cr APL/ Corbis tr NHM 119bl
APL/Corbis br Jesse Fisher tr GI 120tr PL 121br PL tr APL/ Corbis 122bl Jesse
Fisher cr PL 123t GI 124b, c Jesse Fisher t Brian M England 125br,
cr APL/Corbis t Jesse Fisher 126tr APL/Corbis 127br, cr PL cl APL/Corbis
128b, t Brian M England c APL/Corbis 129b, c, t Brian M England 130bl,
tr GI 131bl GI tr PL 132b Jesse Fisher c NHM t Robert R Coenraads 133b
Brian M England cr, tr APL/Corbis 134–135c PD 136c, cr GI cl APL/ Corbis
137c, cl APL/Corbis 138b, tr APL/Corbis 139tl APL/ Corbis 140bl N_EO/Stuart
Snodgrass/Goddard SVS cr GI 141t GI 142bl, t APL/Corbis 143bl PL tr
APL/Corbis 144b APL/Corbis 77tr COR 145br GI 146bl, br APL/Corbis 147bl,
br, tl APL/ Corbis 148bl APL/Corbis cl COR tr GI 149tr PL 150br PL 151bl, tl
APL/Corbis 152bl, cr APL/Corbis 153t GI 154tr GI 155bl Robert R. Coenraads
tl APL/Corbis 156b GI 157cl GI tr APL/ Corbis 158bl, br GI t APL/ Corbis
159br APL/Corbis 160c GI cl PL 162bc PL bl, br APL/ Corbis 163bc GI br PL t
APL/Corbis 164b, tr GI 165bl APL/Corbis 166tr GI 167br GI 168tr PL 169c
GI 170cr GI 171br PL tr APL/ Corbis 172br, tr APL/Corbis 173l AUS/Reg
Morrison 174tr APL/ Corbis 175cl APL/Corbis 176bl Robert R. Coenraads
177cl GI 178tr APL/Corbis 179br PL t APL/ Corbis 180br N_J tr NHM 181cl
APL/Corbis 182cl, cr GI 183bc PD tr APL/Corbis 184–185c PL 186c Brian M
England cl COR cr Jesse Fisher 187c R G Webber cl Brian M England cr GI
188bl, tr Jesse Fisher 190bl Jesse Fisher cr PL 191br, tr Brian M England cr
NHM 192bl AMNH cr Jeff Scovil 193br NHM 194bl, tr Brian M England 195t
Brian M England 196bl Jesse Fisher cr PL 197cl, tl Jesse Fisher 198c Jesse
Fisher cl Brian M England tr PL 199cr APL/Corbis 200tr PL 201br PL
c NHM 202bl, cr APL/Corbis 203cr NHM 204c Jesse Fisher 205bc, bl NHM br
APL/ Corbis tr 206bl, cl Jesse Fisher br Brian M England tr APL/ Corbis
207bl NHM br, tl Brian M England 208b, tr APL/Corbis 209tc, tr Brian M

England tl APL/ Corbis 210bl NHM tr APL/ Corbis 211bl, cl, tl Brian M
England cr NHM 212br, cl Jesse Fisher t Brian M England 213bc Jesse
Fisher br Brian M England 214tr Jim Frazier 215tr Jesse Fisher cr Jim Frazier
tr PL 216b Jim Frazier tr Brian M England 217bl, tr Jesse Fisher br NHM tl
Bill Sechos 218cl Jim Frazier cr APL/Corbis 219bl Jim Frazier br, tr COR tl
NHM 220b Jim Frazier cl R G Webber 221tc, tl, tr Jim Frazier 222c AMNH r
Robert R. Coenraads 223bc PL 224r AMNH 225bl, cl Jesse Fisher r AMNH
226br Jesse Fisher tr AMNH 227br, cr Jesse Fisher l AMNH 228br, c, cr Jesse
Fisher 229bl, cl Jim Frazier br, c, cr Jesse Fisher 230br Jim Frazier cr Jesse
Fisher 231br AMNH cr NHM tr Brian M England 232br AMNH cr Jesse
Fisher tr APL/Corbis 233br, cr, tr AMNH 234b Jesse Fisher c Jim Frazier tc
NHM 235b Jesse Fisher c, t NHM 236cr GI 237bl Robert R. Coenraads cl
Jim Frazier cr APL/Corbis 238–239c PL 240c GI cl APL/Corbis cr PL 241c,
cl APL/Corbis 242b TSA 244bl, tr APL/Corbis 245b APL/ Corbis t NHM 246tr
APL/Corbis 247bl, t AUS/Reg Morrison 248cr GI 249bc APL/Corbis bl PL
250b APL/Corbis tr NHM 251bl NHM 252br APL/Corbis 253tc NHM tl PL
254b APL/Corbis 255bl PL c AUS tl APL/Corbis 256bc, cr GI bl PL tr
APL/Corbis 257t PL 258b APL/Corbis t GI 259cl PL 260bl APL/Corbis br PE
261b PL 262b PL 263br APL/Corbis tl PL tr NHM 264bl, cr GI tr TSA 265bl
APL/Corbis 266br GI tr APL/ Corbis 267bc GI t APL/ Corbis 268bl, cr PL 269bl
GI tl APL/Corbis tr TSA 270bl APL/ Corbis tr NHM 271cl GI 272tr GI 273cl APL/
Corbis 274cr APL/ Corbis 275t APL/ Corbis 276bl, cr APL/Corbis 277cl, tr APL/
Corbis 278b, t PL 279cr, tl PL 280–281c Robert R Coenraads

ILLUSTRATIONS

Andrew Davies/Creative Communication: 23tr, 25b, 43br, 118bl, 188br,
243, 245r, 249r, 251r, 253r, 255r, 259cr, 261cr, 265r, 269r, 275r, 277r
David Kirshner: 45r, 53cl, 54br, 114c
Frank Knight: 55tr, 109tr, 115bl
Map Illustrations & Andrew Davies Creative/Communication: 24b, 56bl,
174b, 205cr, 222bl, 224bl, 226bl
James McKinnon: 90c, 96c, 272b
Peter Schouten: 47cr, 270br, 271b
Wildlife Art Ltd./Richard Bonson: 32b
Wildlife Art Ltd./Robin Bouttell: 70c, 257b, 260t
Wildlife Art Ltd./Brian Edwards: 45tc, 266bl
Wildlife Art Ltd./Cecilia Fitzsimons: 273r
Wildlife Art Ltd./Steve Kirk: 74c, 170bl
Wildlife Art Ltd./Mick Posen: 16cr, 23, 25b, 26b, 27b, 29l, 30br, 31b, 34br,
59tr, 64br, 118bl, 120b, 123b, 126b, 130br, 139r, 141br, 145t, bc, 149b, 150bl,
151r, 153b, 154b, 157cr, 159bl, 160bl, 162cr, 165r, 166bl, 167bl, 168b, 176cr,
177r, 178b, 180bl, 188br, 193tc, 195br, 197r, 198b, 200b, 203tr, 209b, 213tr,
214cl, 221b, 223r, 237t, 246bl, 249cc, 251tl, tc, 252bl
Wildlife Art Ltd./Luis Rey: 115tr
Wildlife Art Ltd./Peter Scott: 275bl
Wildlife Art Ltd./Steve White: 80c, 86c, 102c

ACKNOWLEDGEMENTS

The author would like to thank Brian England and Alex
Ritchie for their assistance in the prepararation of this book.

Captions

page 1 Silvery veils in Ceylon sapphire are the result of tension
within the ordered crystalline structure. Stresses develop
around minerals included within the sapphire as it cooled.
page 2 Sticky resins exuded from conifer trees in a Baltic
forest have trapped a Jurassic-age fly, become buried and
hardened into amber.
page 4–5 Earth viewed from space reveals dunes of red desert
sands, rivers, cloud formations and ocean wave-patterns.
page 6–7 A canyon scene reveals the immensity of geological
time in its horizontal layering.
page 8–9 The rainbow sheen of this Australian opal, which
changes as the stone is moved, is caused by light playing
though arrays of submicroscopic spheres within the stone.
page 12–13 This fossil shell reveals detail of its complex
ornamentation, and even some of the original iridescence.
page 280–2 Surface detail on the octahedral face of a
diamond crystal is enhanced in this false-color image.
Unmistakable are the triangular-shaped corrosion features
that mark the diamond's passage to the surface.